Professor Links and Tinkerton Tinx
Structures

Written by David Jinks
Illustrated by Susan Scurlock

This book is dedicated to all the children, teachers, parents, lecturers, advisers, inspectors and administrators I have worked with over the last forty years. Thank you all for the advice and help on my journey.

David Jinks

Other books in the
The Design Technology series of
Professor Links and Tinkerton Tinx:
Mechanisms
Basic Electrics and Computer Control
Pneumatics and Hydraulics
Workbooks

Also available from EduVision
The Technocat and Mat the Mouse series:
Microsoft® Paint
Microsoft® Paint Workbook
CorelDraw™5
Other computer titles available.

Written by David Jinks
Illustrated by Susan Scurlock
Designed by EduVision Ltd
First published in 2000 by EduVision Ltd

EduVision Ltd is registered in England as Eduvision Limited Registration number 036 19894 Registered Office Mazars Neville Russell. Norwich Union House, High Street, Huddersfield. West Yorkshire. HD1 2LN England.

ISBN 1-903465-00-1
Printed in England

Contents

Introduction

'Jinks Technology', as an essential part of Design & Technology, has a major role to play in developing in children, the lively inquisitive minds needed to cope with the demands of the 21st century. A carefully structured contextual involvement with materials, structures, mechanisms, energy and control (MSMEC) offers all children enjoyment, excitement and expectation as successful designers and makers.

This book is written for all children and teachers wanting to improve their skills, knowledge and understanding of how to build structures. Over the last twenty years, 'Jinks Technology' has evolved in primary school classrooms around the world as more and more children benefited from a system based on materials and tools appropriate to their age and needs. 'Jinks Technology' begins with the recycled materials that primary teachers have used for years to produce a wide, imaginative range of three dimensional structures. It then moves into more resistant materials by giving children the opportunity to create the structures they have designed using 'Lollystick wood' and 'Strip wood'.

'Jinks Technology' - building structural success into primary school Design & Technology.

Just one more packet and I will be able to build a life sized model of the Pyramids!
BIG BISC's
Crackers
CATCHUP
COOKIES
CATS
Crackers
BISC's
COOKIES
CATS
CHOCOLAT
Crackers

Using Recycled Materials

The tradition of using recycled materials to construct things goes back to the 'mists of time' in primary schools. Card boxes and tubes, plus a substantial range of 'throw away' items provide valuable material for appropriate 'Structural' foundations on which to build quality design and make outcomes.

The wide range of construction kits now available has a very important role to play in developing children's early and continuing understanding of how structures are built and how things work. It must be stressed however, that the real value of such kits lies in how far the individual child, or group of children, is challenged into 'thinking and doing' as opposed to the kit providing an experience complete in itself. Together with the traditional range of 'recycled materials', construction kits offer to children the opportunity to tackle well prepared and challenging tasks, thereby developing their skills, knowledge and understanding.

Recycled Materials

Empty cardboard cartons or boxes are really useful for building models. Unfortunately, they usually have lots of pictures and writing on the outside and they are often quite shiny.

The pictures and writing on boxes and cartons can sometimes make them very difficult to use. One idea is to cover the pictures and writing with paint. With some kinds of box, this is the only way to get a 'new' look but you often need a lot of paint. Also, paint and glue doesn't always stick well to the outside of a shiny box.

Most boxes or cartons are actually made out of a flat piece of card called a 'net' which is cut and folded in a special way and than glued together. It is possible to turn these kinds of boxes 'inside out' so the pictures and writing are on the inside and you have plain card on the outside. This technique is called 'box inversion'.

Start by opening out the top and bottom of your box. Look carefully inside and you will usually find a narrow flap or tab of card where it is glued together. If you are careful, you can pull this tab away from the box and open it out. Don't worry too much if it tears a little. The tab will be hidden again inside your box when you invert it.

Once your box is opened out, it can be re-folded 'inside out' so the pictures and writing are now on the inside.

The shape that makes your box is called a 'net'. Depending upon the box you have used the 'nets' will all look slightly different. They will all, however, look something like the one on the right. Look at the 'nets' for lots of different boxes and try to work out what all the different cuts and folds are used for.

Now, fold the box 'inside out' so the pictures and writing are now on the inside. Make sure that you put sharp creases on all the folds (the other way round). Apply some PVA glue to the tab and use it to glue the box back together again. Please remember it takes time for the glue to set.

You should now have a box with plain card sides which can be painted and glued to suit your own designs. Boxes of different shapes and sizes may then be glued together to form interesting finished structures.

If you want a particular shaped structure e.g. a train, you can often create the shape by joining several boxes together. The boxes can be very firmly joined using PVA glue.

Sometimes you will think of a finished design and have to find boxes of the correct shape and size.

Sometimes the shape of the boxes you have can trigger ideas for the final design of the model.

You can then glue the boxes together and then you need to find all the materials to add detail to your final design. Here, card discs, triangles, lollysticks and bottle tops have been used.

You can also paint extra details onto your model and even cut away some of the card to make windows, doors etc.

It is often much easier, however, to paint and cut the card **before** the boxes are glued back together as you will then be working on a flat surface.

You will often want to add extra detail to your models. It is often much easier to plan this before you start making. You can then add the detail while the boxes are still flat.

First, open out your box so that all the sides are flat.

Next, work out which part of the net will be the front, back, top, side etc. of your finished model.

Then, paint your box and add all the other details that you want.

Finally, fold your box back together and glue the tab. Add any other details such as the painted card discs used as wheels for this bus.

You will often need to cut out shapes or make folds in the card. You can sometimes use scissors for this but if you need to cut in the middle of the card, as for the windows shown below, you will have to use a craft knife and metal safety rule.

Young children should **never be allowed to use a craft knife** on their own. The teacher (adult) must do this part of the work. The diagram below shows the net of a box which will eventually become a house.

The door has been hinged by cutting along three sides and folding the fourth side. The best way to get a neat fold is to use an old biro that has run out of ink. Put your ruler along the line where you need to produce your fold and draw along the line with the empty biro. The indent it leaves will allow you to make a very neat fold. An old bottle top has been glued on for a handle.

The 'Olfa' Compass Cutter shown on this page will cut circles in paper and card. It is used in a similar manner to an ordinary compass but has a sharp blade rather than a piece of graphite. The Compass Cutter should always be used on a cutting mat thereby protecting the cutting blade and the surface on which you are working. Do not use undue force, it is better to rotate the cutter a number of times to produce the circle you want. The cutter will only cut in the direction the blade is facing. The blade can be removed and turned round to accommodate left handed users. The cutter should never be used by young children. It is a 'teacher tool'.

Here the compass cutter has been used to cut holes from the flat net of a box. The holes have been made the same size as some card tubes (toilet rolls, kitchen rolls etc.).

When the net is inverted and re-assembled, the holes allow you to attach the tubes more easily by pushing them through the holes and PVA gluing them in place.

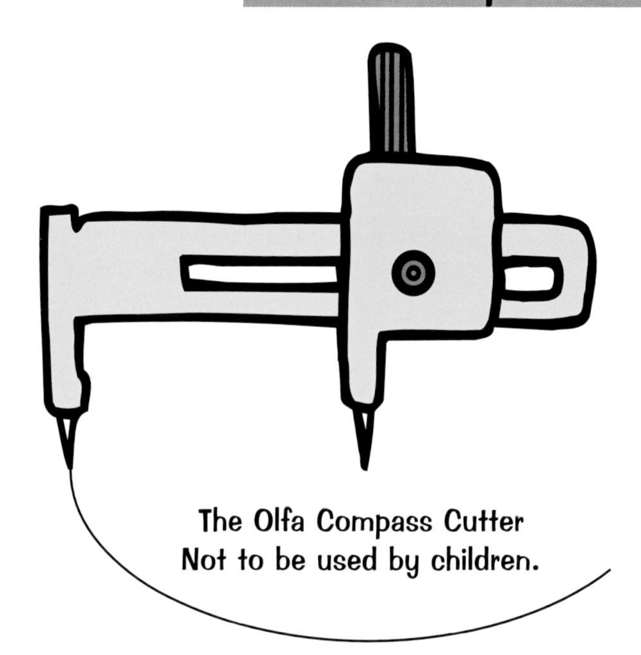

The Olfa Compass Cutter Not to be used by children.

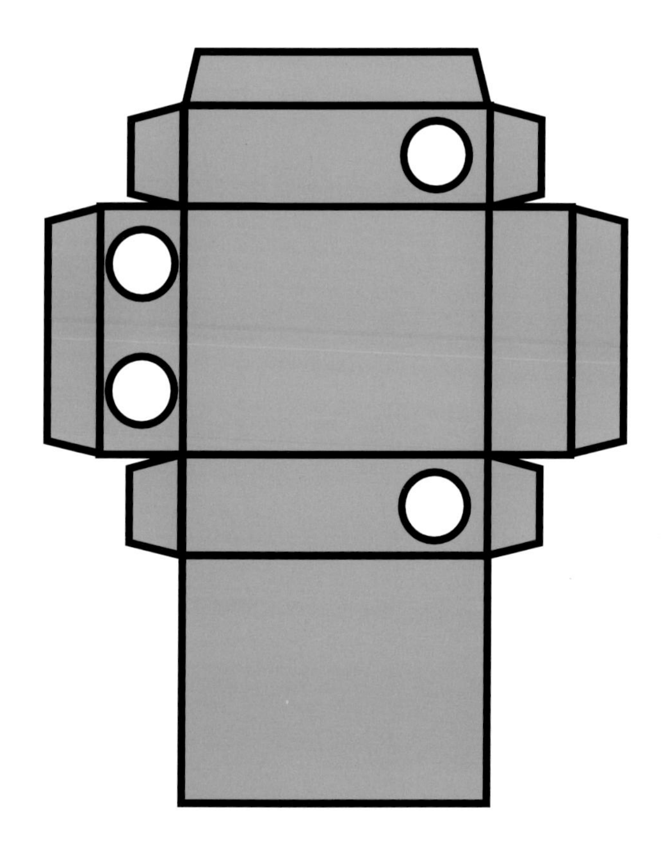

Card tubes pushed through the holes. Apply PVA glue wherever the tubes touch the box.

Hinges

A hinge is a type of Lever. It allows two things to be joined together (eg door and frame) but offers one of them movement (the door opens and closes, the frame does not move). There will be occasions when you will use hinges in your designing and making.

As usual start by choosing the boxes you need for your final design. Here, three boxes have been chosen for a tipper lorry.

'Invert' your boxes and add any other details that you want for the cab etc. before you put the boxes back together. In this example the cab and bed of the lorry are first glued together with PVA adhesive. The back of the lorry which will eventually tip is an open box.

As you can see, the hinge allows the open box to 'tip' up and down.

Other details like the card discs used here for wheels are then added. There are many ways to make your hinge. The next page gives three simple ideas but there are many other possibilities.

a) 'Correx' or 'Corriflute' Hinge

'Correx' or 'Corriflute' is a type of plastic material similar to corrugated paper. The side view above shows how it is made. If one side of the 'fluting' is cut along its complete length it makes an excellent and very strong hinge. Unfortunately, the plastic material does not stick very well with PVA glue so your teacher will have to use a hot glue gun or impact adhesive.

b) Taped Hinge

A taped hinge is the easiest type to use. Simply, cut a piece of sticky tape (sellotape, masking tape, Copydex tape etc.) to size and lay it over both boxes. The stronger the tape you use, the stronger the hinge.

c) Folded Card Hinge

A folded card hinge is also quite easy to make. Cut a piece of card to size so that it is as wide as your boxes and mark and score a fold line. A very neat and easy way to do this is to use an empty biro pen with a straight-edged rule. Make a neat, sharp fold and then open the card back out. Glue the hinge into place over the two boxes with PVA adhesive making sure that the fold line lies between the two boxes.

Axle Strengthener

A single hole punch or a rotating head punch are used to make very neat holes in paper and card. However both these tools are unable to punch holes in wide pieces of paper or card. One way around this problem is shown here.

First, mark the position of your hole and then make the hole with a sharp pencil, the hole should be slightly bigger than the hole made by a punch.

Next, cut a square or rectangular piece of thick card. The easiest way to mark the centre is to join the two diagonals (corner to corner). Using your single hole punch, punch a neat hole where the two lines cross.

You have now made a 'hole or axle strengthener'. Glue the strengthener in place over the hole with PVA adhesive. The axle strengthener does two jobs. It makes the hole much neater, allowing 5mm axles to rotate freely and it also strengthens the card at the same time.

Mark your box front and back so the holes will align then use a sharp pencil to make the holes.

Glue two axle strengtheners in place over the holes you have made with PVA adhesive and leave to dry.

Here, this technique is being used on a box. The box could become a windmill.

5mm wooden dowel has many uses in the 'Jinks Technology' system. One of the major uses, however, is for axles on wheeled devices/vehicles.

The same principle applies when any piece of wood has to be attached to an axle e.g. a wooden winder etc. If you require the wheel to be a 'loose' fit on the axle then use a 5mm drill.

Since the 5mm axle has to be forced into a 5mm hole, it is very useful to taper the end of the axle very slightly. An ordinary pencil sharpener can be used to do this. Remember, however, to make a taper, not a point! To stop the wheels rubbing on the box or frame of the vehicle, use a small piece of 5mm PVC tubing as a spacer. This spacer cuts down the friction between the wheel and the vehicle chassis.

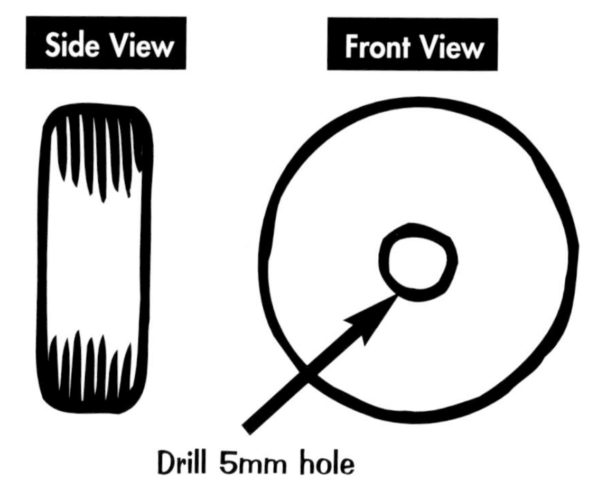

Less Friction

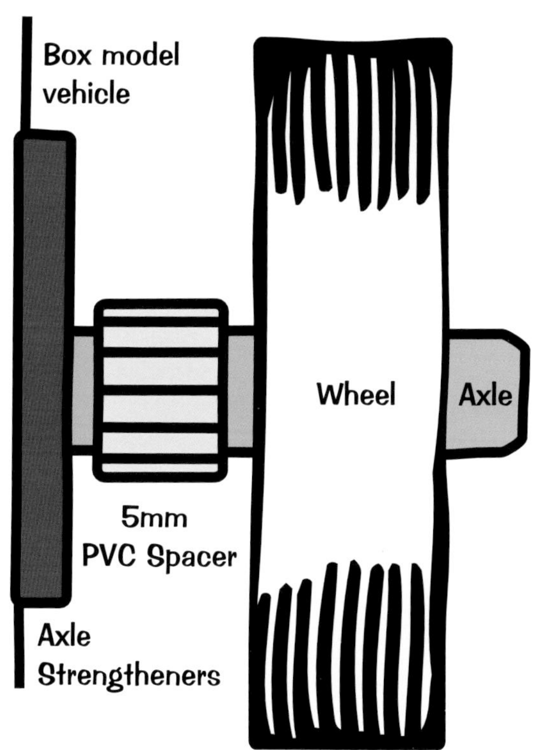

The **taper** helps you force the wheel onto the **axle**.
It also helps to *gently* **twist** the wheel as you push.

Axle Holes

Different shaped boxes can be made into all kinds of vehicles but to make them 'really move' you will have to add axles and wheels which can actually turn.

Quick holes for axles and wheels.

An easy way to make holes in small card boxes and cartons is described here. It relies on the fact that most boxes have opening flaps at the top and bottom. Once opened out in this way, the boxes become much less rigid and can be folded over flat to either the left or the right. Indeed, this is an excellent way to store boxes as they take up much less space in this form. A single hole punch is then used to punch neat holes to carry axles and wheels.

Fold the box to one side as shown opposite and use the punch to cut the holes marked 1 and 2. Because of the way that the box has been folded, you will actually be cutting through a double thickness of card. Try to ensure that the holes are evenly spaced and the same distance up from edge A. Fold the box the other way around and holes 3 and 4, in the sides of the folded box will be visible.

Now repeat the process outlined on the previous page but this time use your punch to cut holes 5 and 6 along edge B.

Again, try to ensure that the holes are aligned with holes 1 and 2. This will be important to the proper running of the finished 'vehicle'.

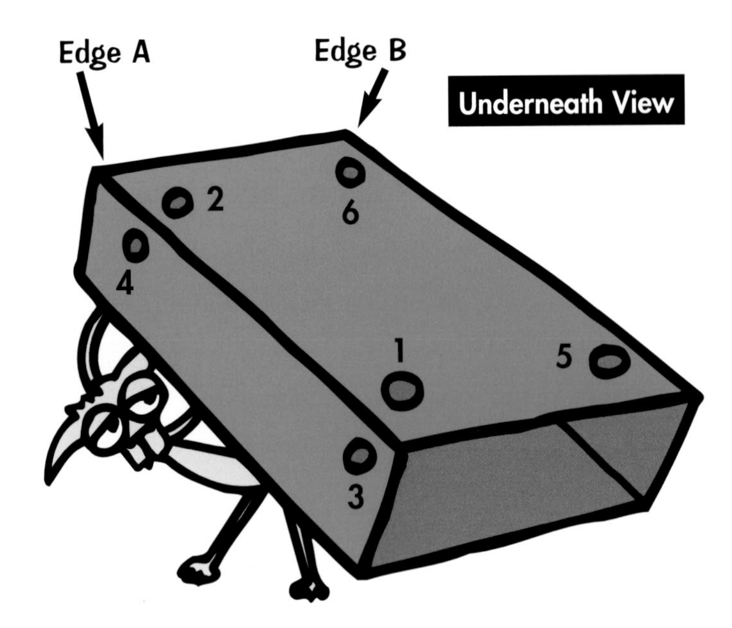

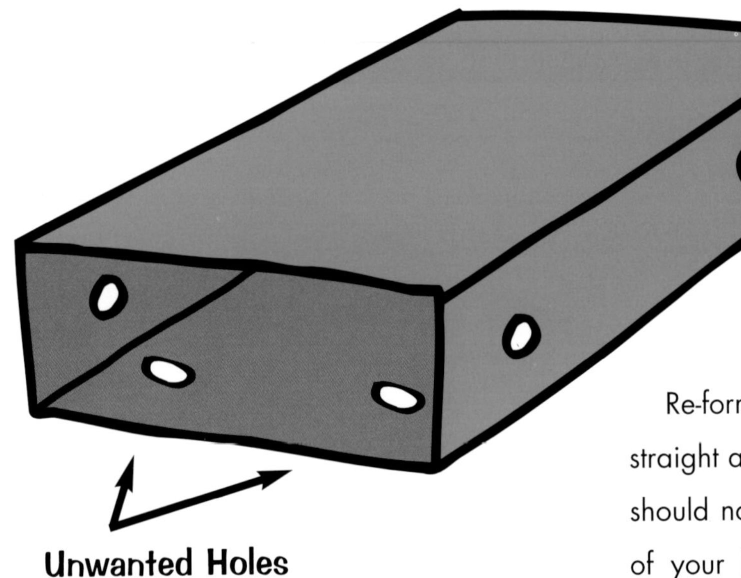

Re-form your box by folding it up straight and tucking in the end flaps. You should now have neat holes in the sides of your box, ready to take axles and wheels. You will, of course, have four unwanted holes in the bottom of your box but these will not be seen when your vehicle is viewed from above. The vehicle is finished by adding axles, wheels, spacers, extra detail etc.

It may be more appropriate to use ready-made wheels and cut axles. Also remember that axles are used with other mechanisms e.g. pulleys and gears.

The **unwanted** holes will be hidden *under* the vehicle.

Vehicles - An Oil Tanker

Different shaped boxes can be made into all kinds of vehicles but to make them 'really move' you will have to add axles and wheels which rotate.

Measure and mark the four points where the 5 mm wooden axles will eventually go through the chassis. Use the point of a pencil or something similar to make holes at these positions. The holes should be bigger than the axles. Paint your box and put in other details if you want to. Glue the box back together and join it to your other boxes.

Make four rectangular axle strengtheners and glue them over the large holes in your model. Cut four rectangular pieces of thick card and draw lines corner to corner. The lines meet at the centre of the rectangle. Use a punch to cut neat holes at this point on your strengtheners.Measure and cut two axles. Get four wheels and PVC spacers and put them together. Your vehicle should now move very easily.

Choose your boxes and decide which one is going to be used for the base or 'chassis' of your vehicle.

The PVC spacers will make the vehicle move more easily.

Vehicles - A police Car

Another way to mount axles and wheels is to cut thick card, extended axle holders and glue them to your box model with PVA glue.

Choose your boxes and add all the detail you need for your finished design. Glue the boxes together to make your vehicle.

Cut four extended axle holders from thick card.

Measure and mark some pencil guidelines onto your model so that you will know where to stick the axle holders.

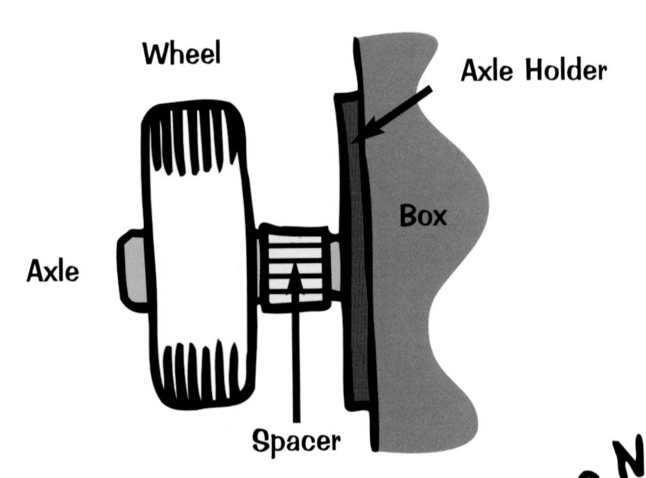

Remember that the two holders which will carry one axle have to be aligned accurately or your vehicle will not run straight.

Glue the four axle holders into position with PVA glue and leave them until the glue has set.

Finally, fix your four wheels, four spacers and two axles to your vehicle.

The technique of 'lollystick wood' comes from the way Roll-top desks were made. Then they would glue strips of wood to a leather backing which, when finished, allowed a seemingly solid piece of wood to 'roll' around a curved frame. If you use lollysticks instead of wood and thin card instead of leather the results are amazing. The diagrams illustrate how to make this flexible type of wood. You can make many things using lollystick wood.

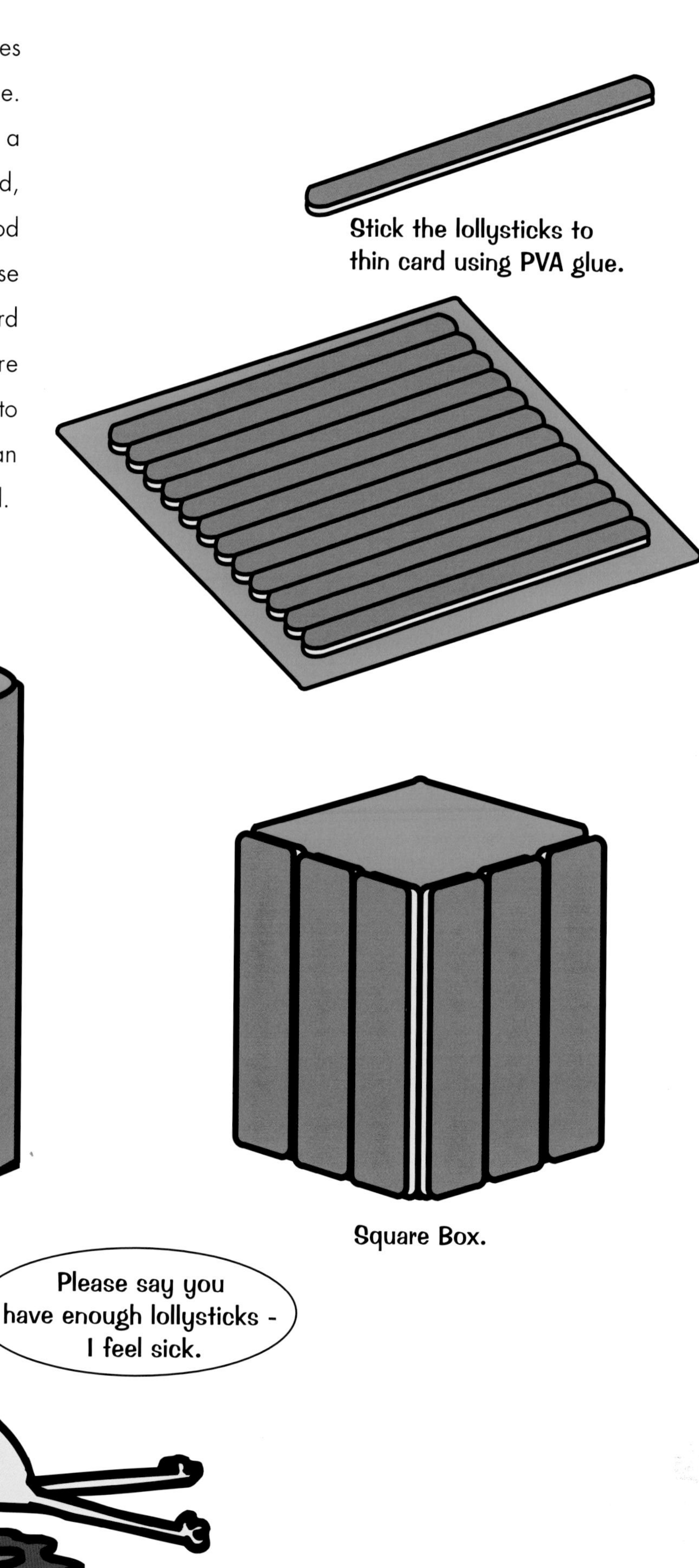

Is this what they call a birds-eye view?

Building Structures

'Jinks Technology' uses recycled materials and two sizes of softwood strip, 8mm x 8mm and 21mm x 8mm, to build most structures. Lollysticks and thin card are also used to make 'Lollystick wood': a most useful structural material. The same technique can be used to change 8mm square and 21mm x 8 mm individual pieces into solid 'strip wood'.

A structure is something that will support its own weight and whatever weight or load is put on it. A chair is a good example of a manufactured structure.

The chair has to be strong enough to support its own weight and the weight of a person sitting on it.

There are two different types of structure, a 'frame' structure and a 'shell' structure.

Frame structures are made from strips of material joined together to form a framework e.g. a crane, a pylon, a suspension bridge.

Shell structures are held together by the skin of the material that they are made from e.g. a drinks can, a milk carton, a suitcase.

Very often a structure consists of both a frame and a shell structure e.g. our bodies. The skeleton of our bodies is a frame structure and this is covered by our skin which is a shell structure. Together, they make our bodies very strong.

Shell and Frame Structure

A structure must be able to support the weight put on it or it will collapse. This is called structural failure. Structures can be strengthened in many ways to prevent them collapsing. This is called stabilising a structure.

In the example shown, the chair collapses as Tinkerton sits on it. This is called structural failure. Since Tinkerton's weight is causing the legs to push apart, this structure can be stabilised easily by adding cross supports which will hold the legs together.

The triangle is a very strong shape and it is often used to stabilise structures. You can experiment with this by cutting yourself some strips of thick card and joining them with paper fasteners as shown below. The rectangular frame changes shape very easily. Now add another piece of card to form two triangles. Try to change the shape now. What do you notice?

A structural failure

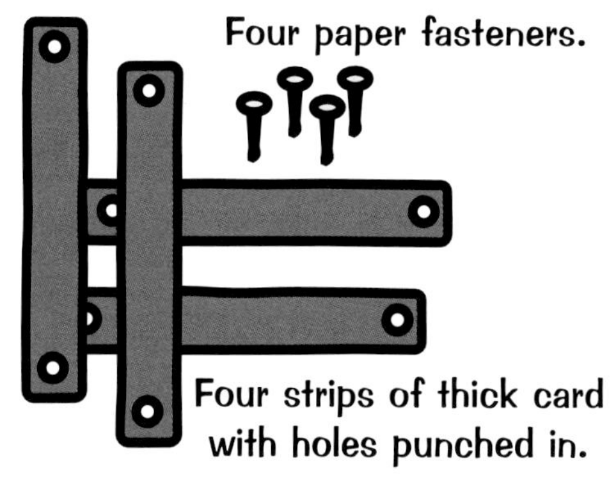

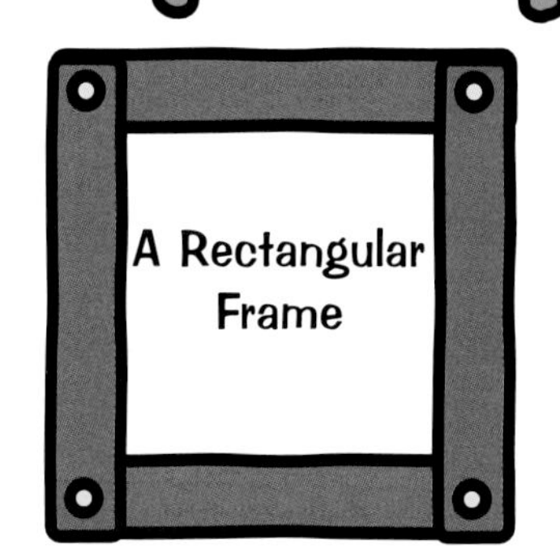

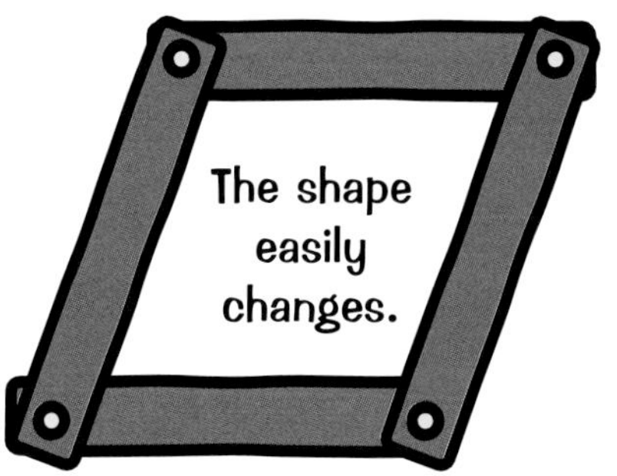

Frames

You can make your own frames and structures using square section wood and thin card, constructional triangles. Using this method you can design and make almost any kind of structure.

In 'Jinks Technology' two sizes of softwood are used to build the majority of structures.

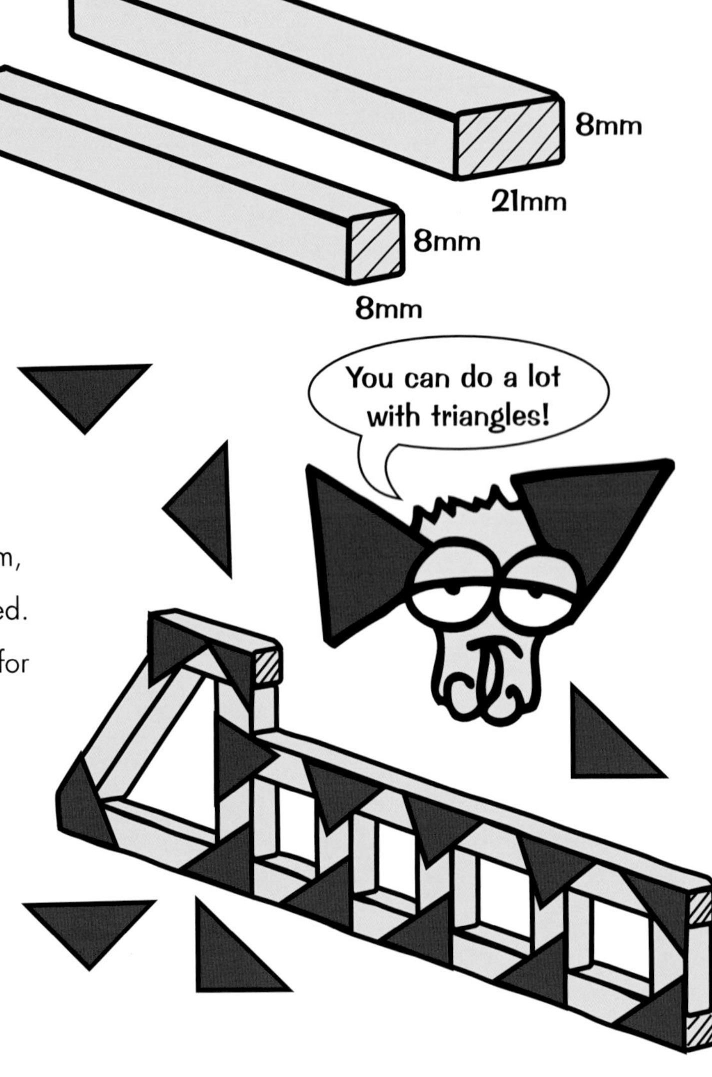

The main size for building frames is 8mm x 8mm, 21mm x 8mm gives added strength when needed. 21mm x 8mm is also used if there is a need for holes to be drilled in the structure.

Whichever size you use, it can be made into a frame by gluing it together with PVA adhesive and strengthening the corners with thin card constructional triangles. These flat frames can then be joined together to make almost any kind of structure.

The example here shows the body of a lorry. It has been made by joining two, identical, shaped frames. All that is needed to complete the structure is a set of axles and wheels and a card body.

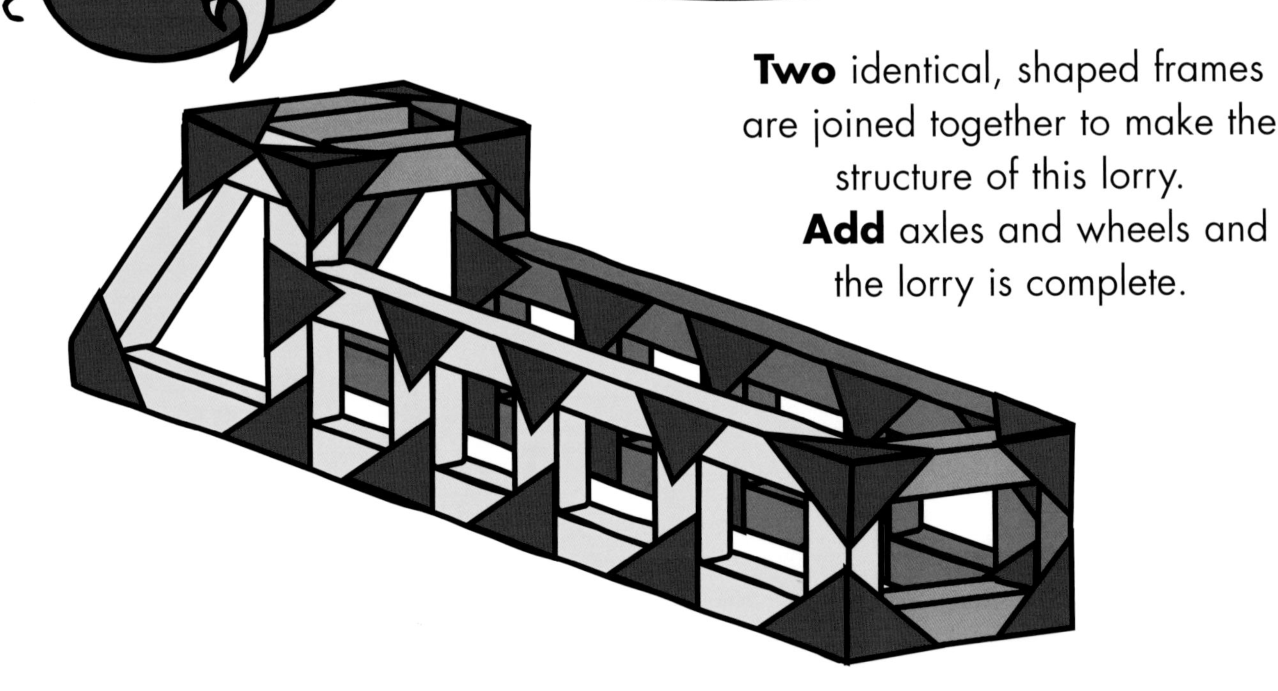

Two identical, shaped frames are joined together to make the structure of this lorry.
Add axles and wheels and the lorry is complete.

You will need to make lots of different triangles and other shapes to build your designs. They can all be made by starting with a grid of squares drawn onto thin card.

You start with a piece of thin or thick card.

Thin card is used for constructional triangles; thick card is used for axle holders and strengtheners.

Place your ruler along one edge of the card. Draw a thin, firm, pencil line.

Now, place your ruler on the pencil line that you have drawn and draw another parallel to the first line.

Repeat this until you have lots of lines, all a ruler's width apart (approximately 3 cm).

Turn your card through 90 degrees and repeat this process, making lots of lines in the other direction.

You should now have a sheet of card with a grid or pattern of equal squares that are the same width as your ruler (approximately 3 cm).

Approximately 3cm

Constructional Triangles

To build your designs you will have to join pieces of square section wood to make frames and structures. This is done with constructional triangles.

Constructional triangles are made in **Thin** card. You start by making a pattern or grid of squares.

Now join up the diagonals (corner to corner) of the squares using a ruler and pencil.

For single triangles, join them as shown in **figure 1**.

For double triangles, join them as shown in **figure 2**.

Carefully cut off strips of triangles from your card. Finally, cut off the individual single or double size triangles.

Single constructional triangles are for making joints and building flat frames.

Double size triangles, when folded down their centre line allow you to build up from the corners of flat frames to make three dimensional structures.

If you use thick card instead of thin, you can make lots of different shaped axle holders and strengtheners. A thick, double-sized triangle makes an axle holder which can be used for attaching wheels, pulleys, winches, winders, ratchets, gears and belt drives to your frames and structures.

An 'L' joint allows two pieces of wood to be joined at right angles. 'L' joints are used to make the corners of rectangular frames.

First, mark out and cut to size the two pieces of wood to be joined. Always try to cut the ends of the wood straight.

Next, put a small amount of PVA glue onto one piece and press it against the other piece.

Take a constructional triangle and cover one side with PVA glue and press the triangle firmly into place at the corner of the joint.

Carefully, (or it will fall apart) turn the wood over and apply a second triangle to the other side of the joint.

Make sure that the two pieces of wood are at right angles and that the triangles are firmly pressed down onto the wood. Carefully remove any excess glue and leave to dry.

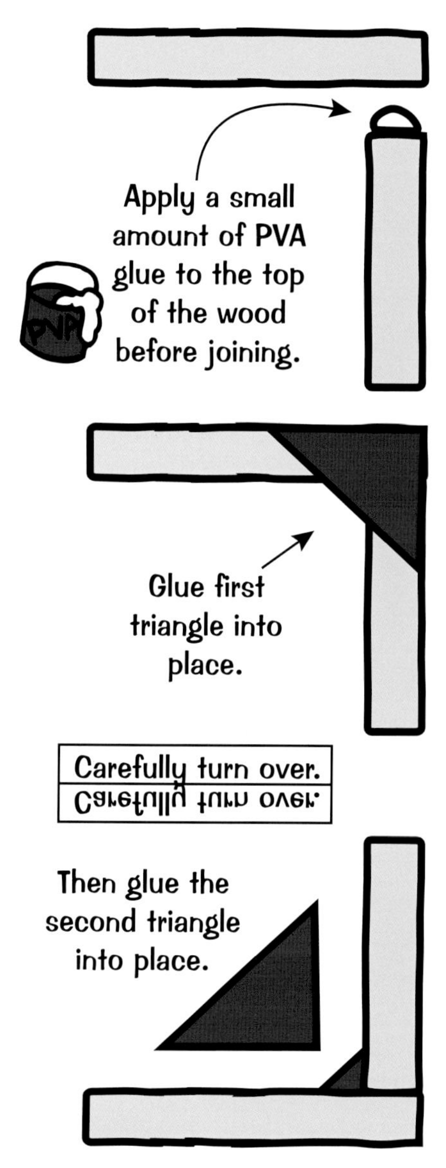

Joints - A 'T' Joint

A 'T' joint allows one piece of wood to be joined to another piece anywhere along its length.

First, mark out and cut to size the two pieces of wood to be joined. Always try to cut the ends of the wood straight.

Next, put a small amount of PVA glue onto one piece and press it against the other piece.

Take a constructional triangle and cover one side with PVA glue and press the triangle firmly into place as shown in step 2.

Carefully, (or it will fall apart) turn the wood over and apply a second triangle to the other side of the joint.

Make sure that the two pieces of wood are at right angles and that the triangles are firmly pressed down onto the wood. Remove any excess glue and leave to dry.

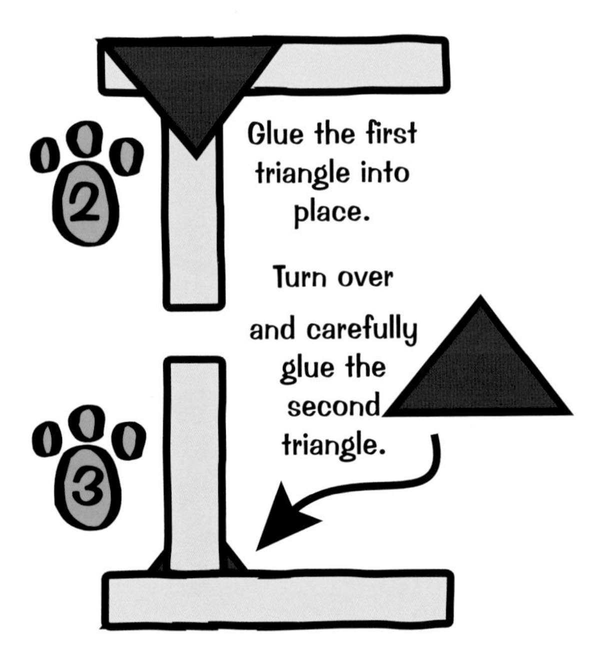

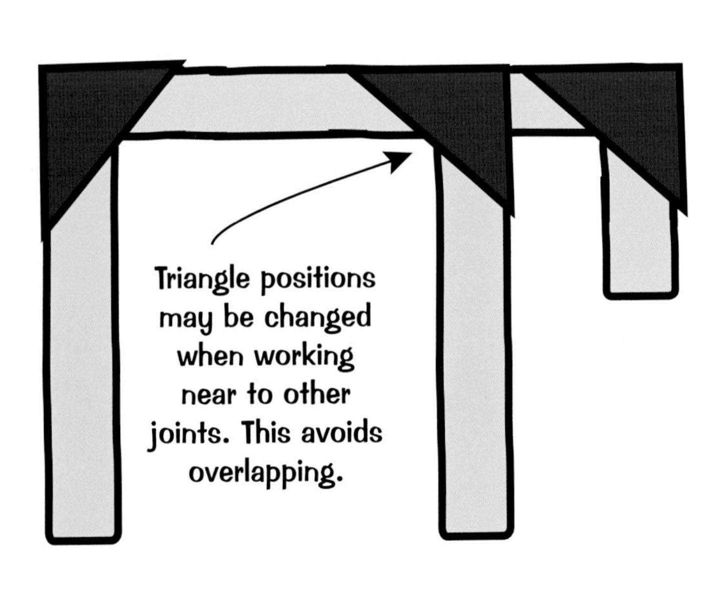

An 'Angled' joint allows one piece of wood to be joined to another piece at any angle.

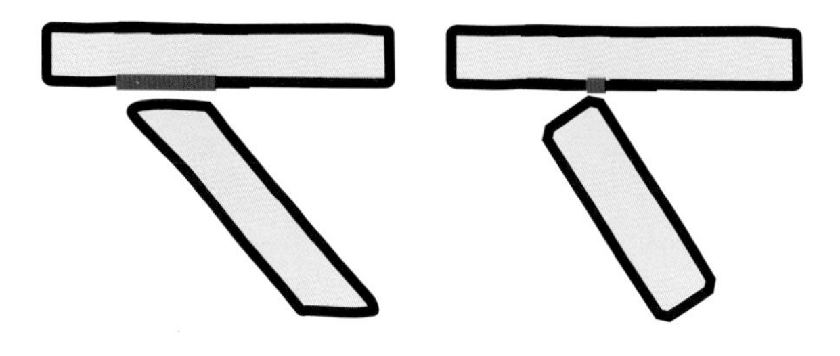

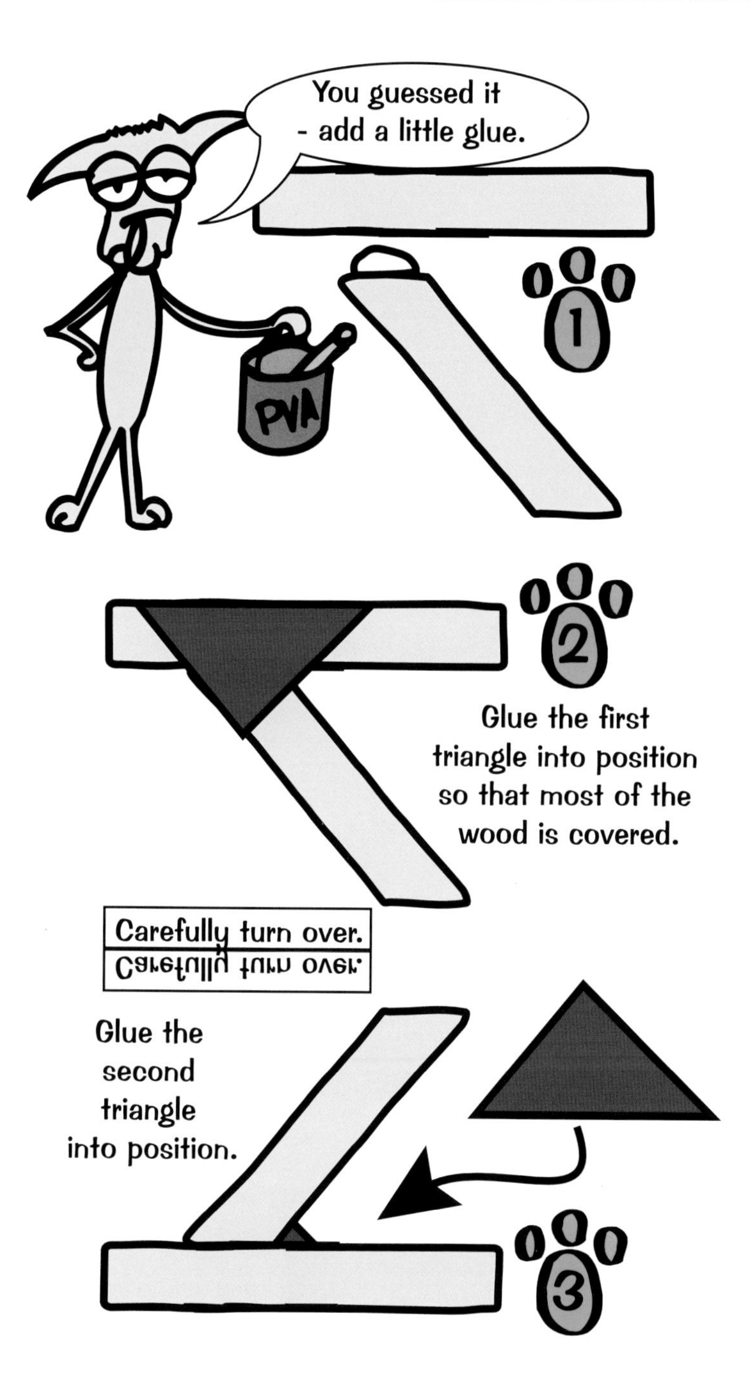

First, mark out and cut to size the two pieces of wood to be joined. Try to mark and cut the wood at the correct angle so there is a lot of wood contact.

Next, put a small amount of PVA glue onto one piece and press it against the other piece. Take a constructional triangle and cover one side with PVA glue and press the triangle firmly into place as shown in step 2. Turn over and glue the second triangle into place then leave to dry.

The position of the triangle will vary depending on the angle required. Always try to get as much card in contact with the wood as you can.

Frames - 2 Dimensions

Two dimensional frames made from square section wood will strengthen your box models. Frames can be made in a variety of shapes but the rectangular frame shown below is the most common.

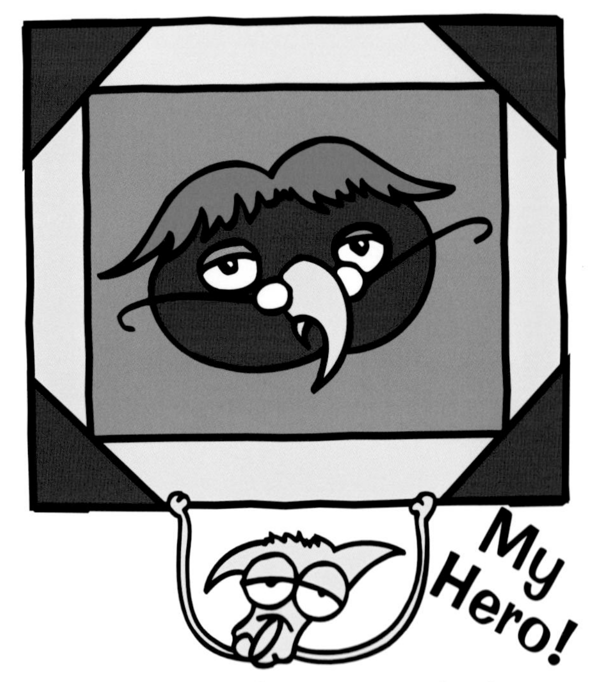

Four L Joints make a rectangular frame.

A Rectangular Frame

First, mark out and cut to size the four pieces of wood to be joined. Always try to cut the ends of the wood straight. The frame is then assembled using four 'L' joints (8 constructional triangles).

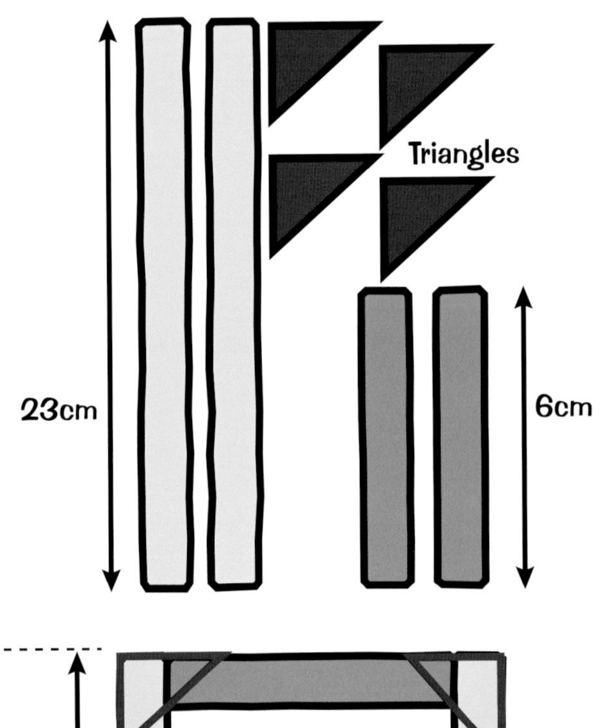

Take care to assemble the four pieces of wood as planned since a different assembly can change the shape of the frame. In the example shown below, two 23cm and two 6cm lengths of square section wood have been joined to make a rectangular frame.

As you can see, however, because of the way they have been put together, the end result can be very different.

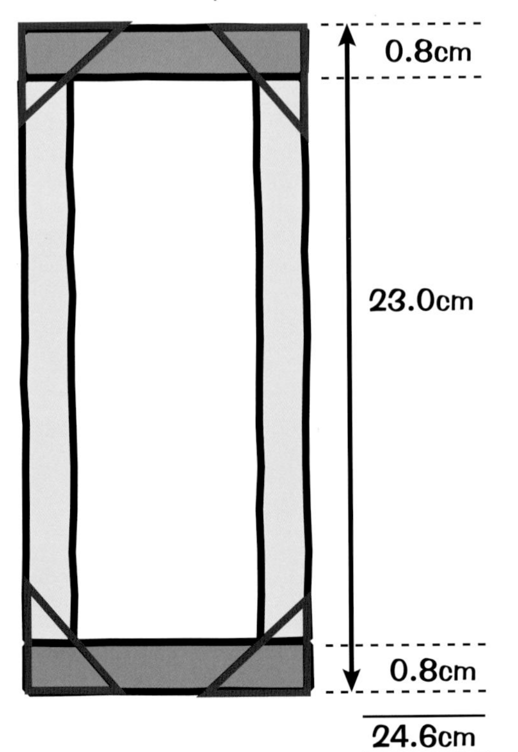

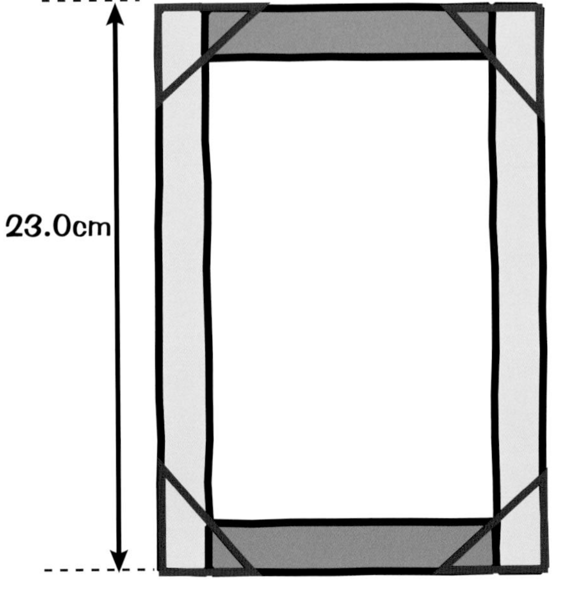

Two-dimensional frames are the starting point for any structure. Frames can be made in a variety of shapes e.g. a rectangular, triangular, 'T' shape etc.

An 'A' Frame

An 'A' frame is also made up from four pieces of wood but the 'A' frame is best assembled using two 'L' joints and two 'T' joints.

'A' frames have a variety of uses but are particularly suitable for three wheeled vehicles.

The diagram opposite shows the eventual position of a 'cotton reel' roller and two wheels.

An A Frame

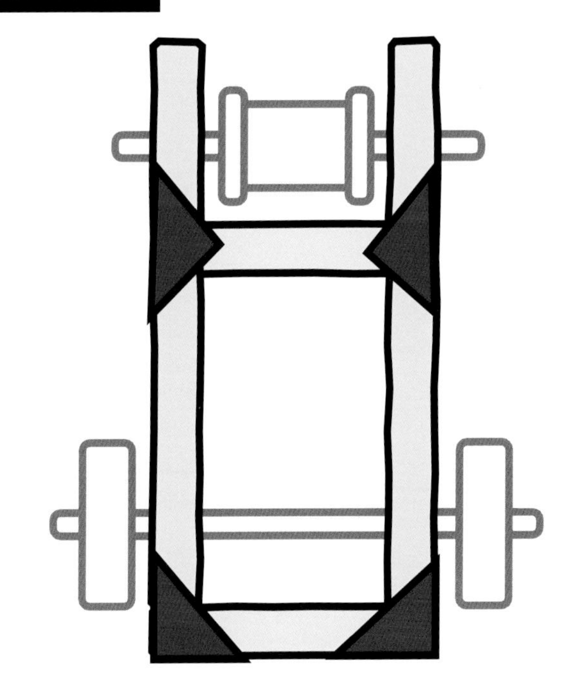

A Triangular Frame

A triangular frame can also be used for three wheeled vehicles. Care should be taken to cut the longer pieces of wood at the correct angle to allow the wood to be glued to the cross piece. It is often useful to plan this kind of frame on squared or graph paper. The cutting angles can then be transferred to the wood to be cut.

To produce a neat finish, constructional triangles and a square of card are first overlapped over the frame. When the glue has dried, the excess card can then be removed with a pair of scissors.

A Triangular Frame

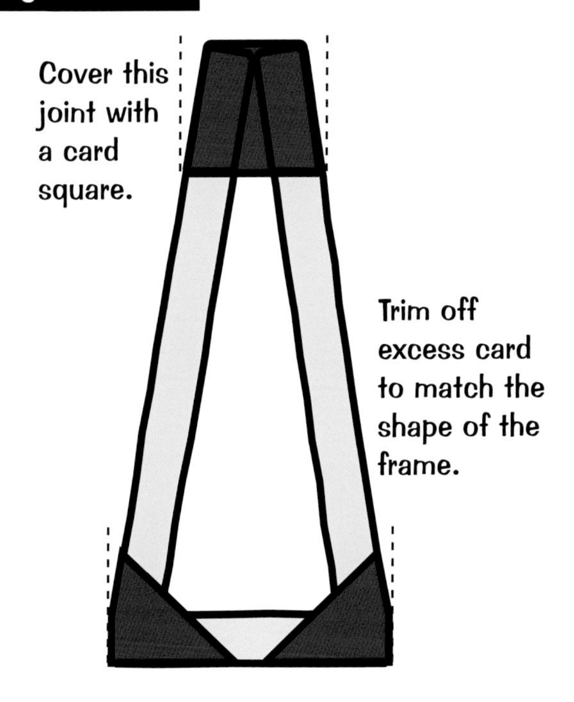

Shaped Frames

Shaped frames make your designs look more realistic. Two identical frames are joined together to form a structure e.g. a lorry. Shaped frames should be planned out very carefully before the wood is cut.

Try to keep the shape of your frame as simple as possible since you will have to make two identical frames and join them to make your structure. Strengthen large frames with cross supports.

Use your plan as a cutting guide and carefully mark out and saw all the straight cuts. Since you will eventually need two identical frames, remember to cut two of everything. Use constructional triangles to join and strengthen your frame.

Cut two of every piece.

Join the wood with PVA glue and constructional triangles.

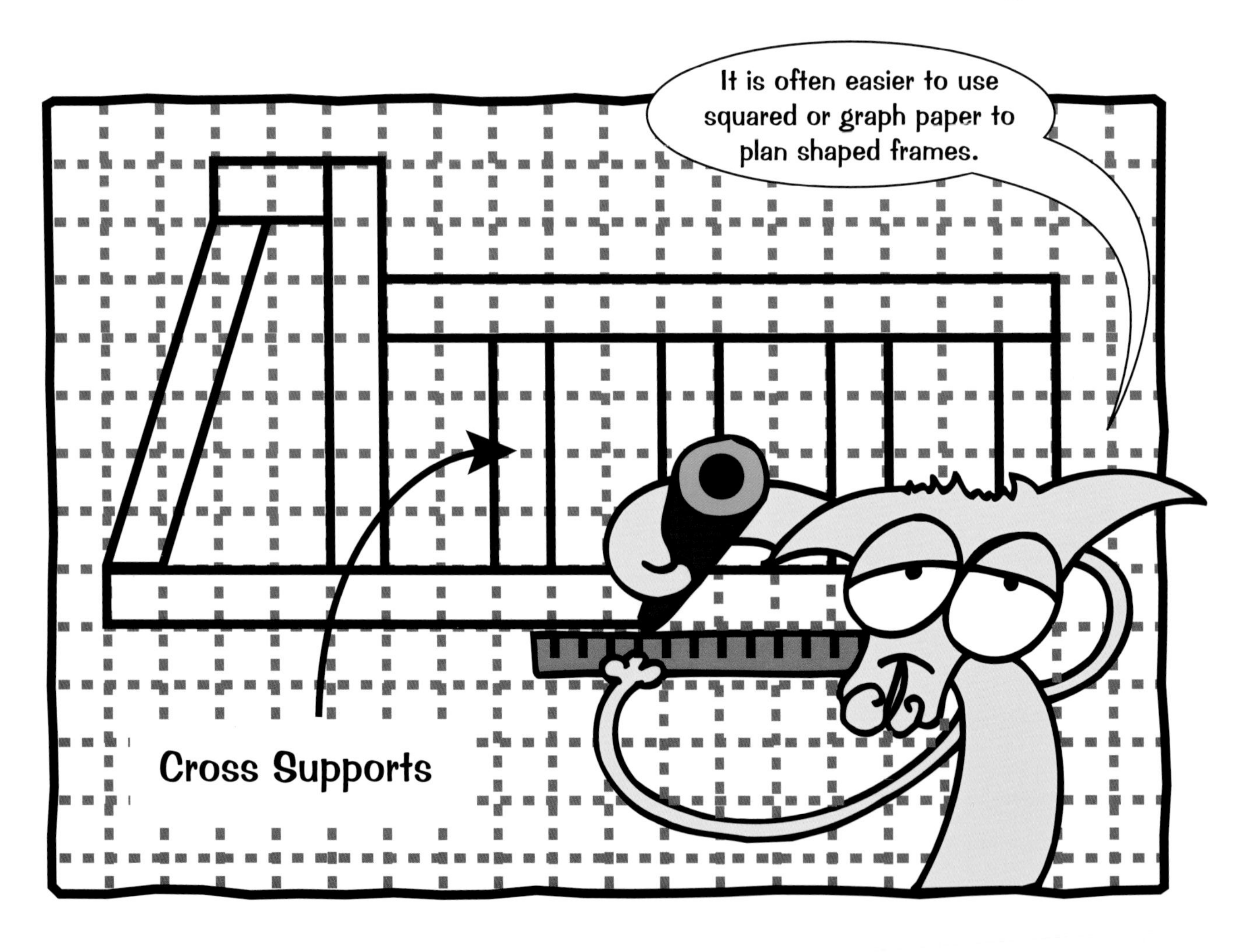

Lay the frame over another piece of square section wood and mark the angled cuts with a sharp pencil.

The wood can then be marked for the angled joints. Lay your frame over a piece of wood and line it up correctly. Use a sharp pencil to transfer the exact cutting angles to the piece of wood then saw to size. Finish the frame with more constructional triangles trimmed to size.

For your finished structure you will need to make another identical frame. The two shaped frames will eventually be joined with several cross supports of the same length.

Since you will usually have to make two identical frames it is very important that you keep your designs as simple as possible. Try to have as few angled cuts as possible since these are harder to duplicate. With imagination, even simple frame designs can produce very impressive and realistic structures.

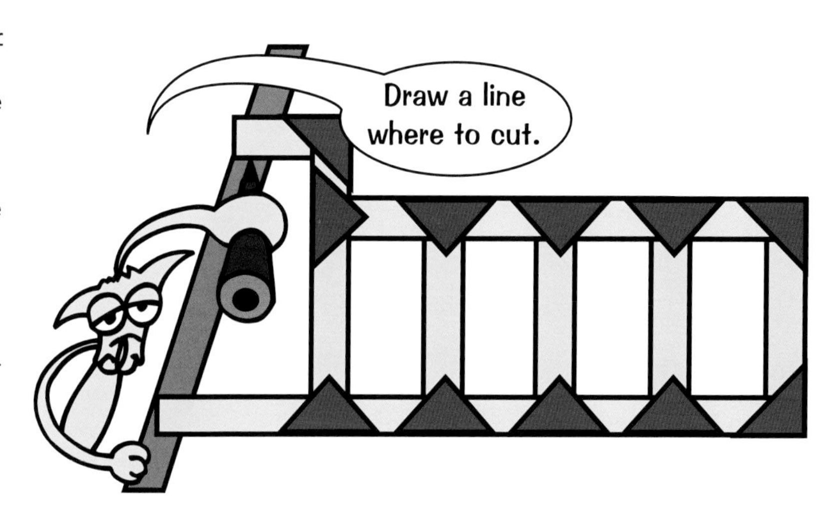

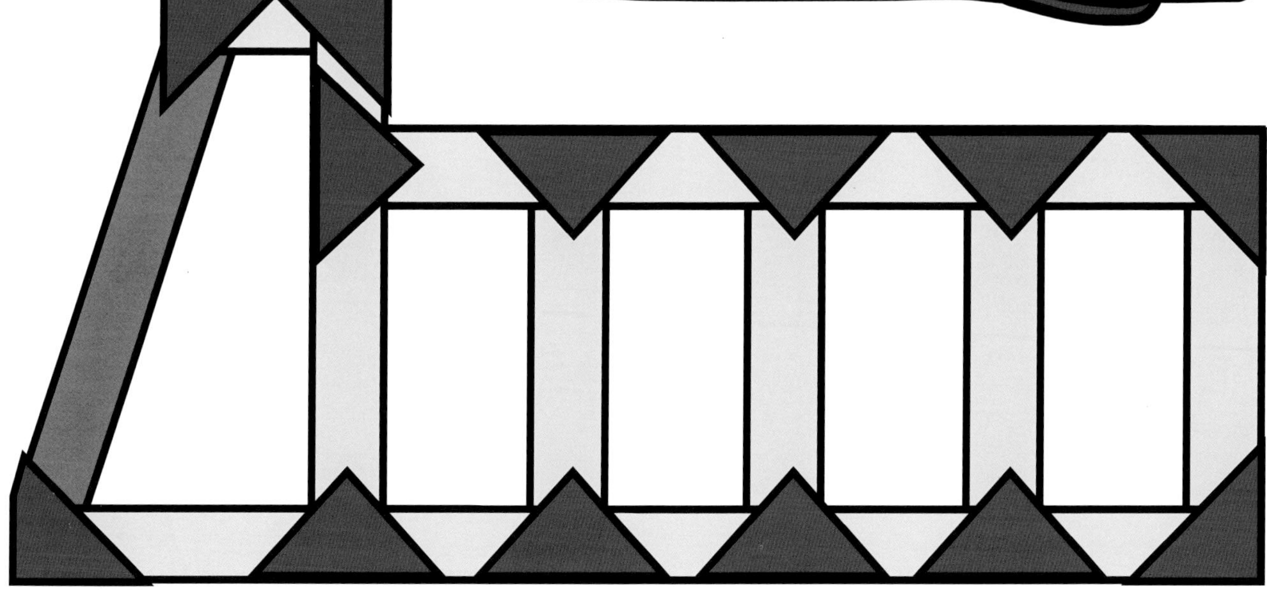

Four Shaped Frames

The wood shown below and on the next page has been cut to make four different shaped frames. The finished frames are also shown. You can see from the wood that the designs are very simple. There are only one or two angled cuts in each case.

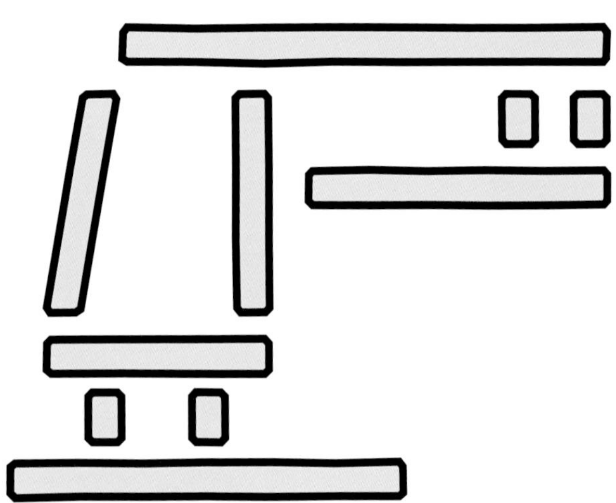

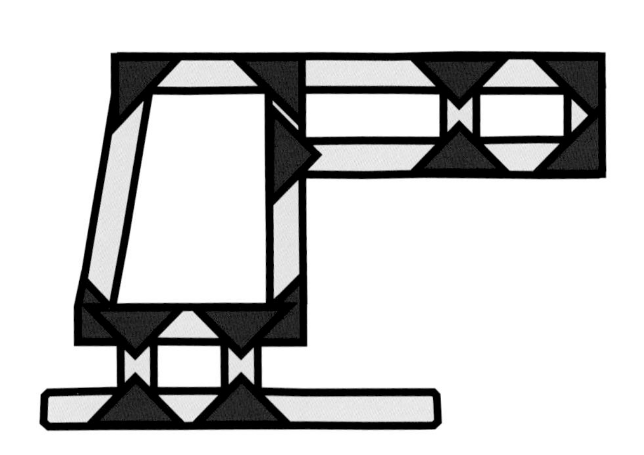

A Helicopter

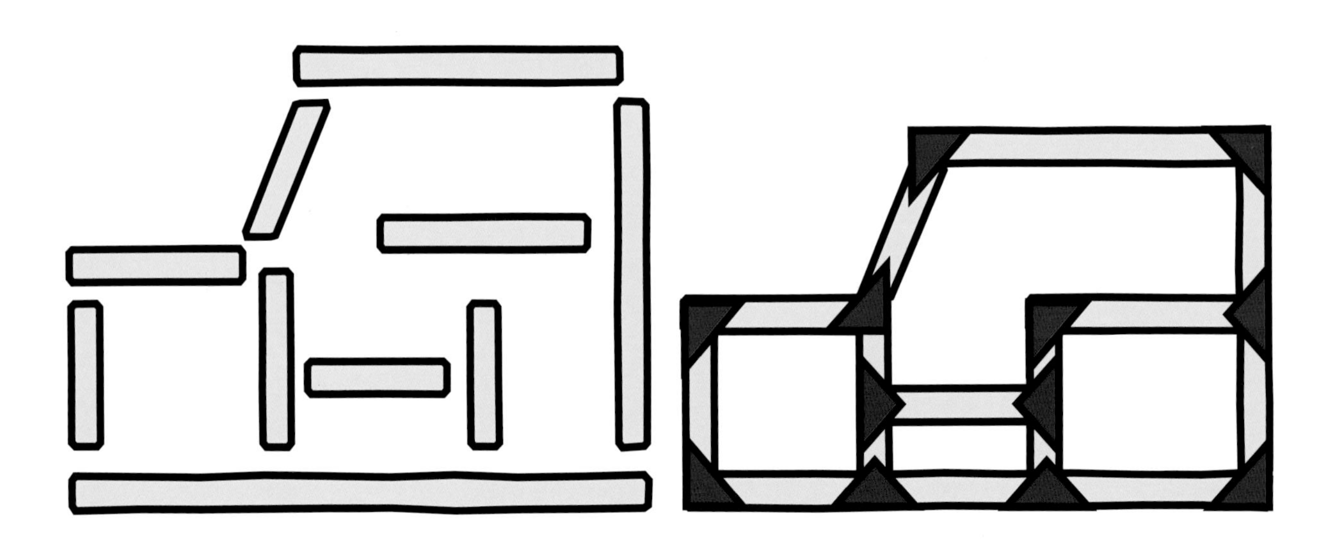

A Car

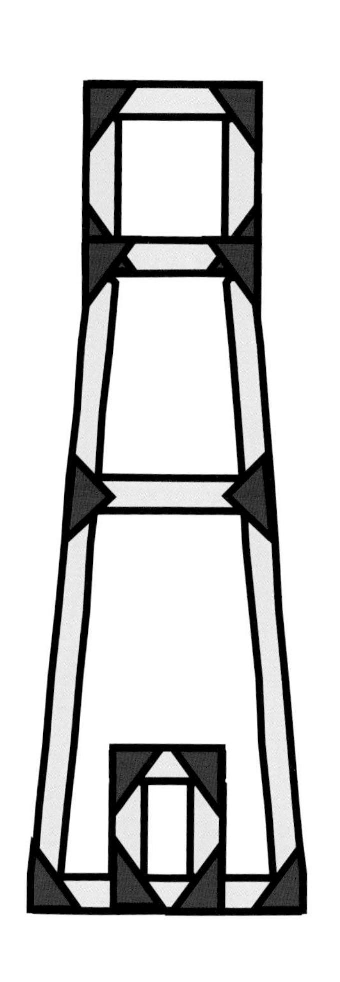
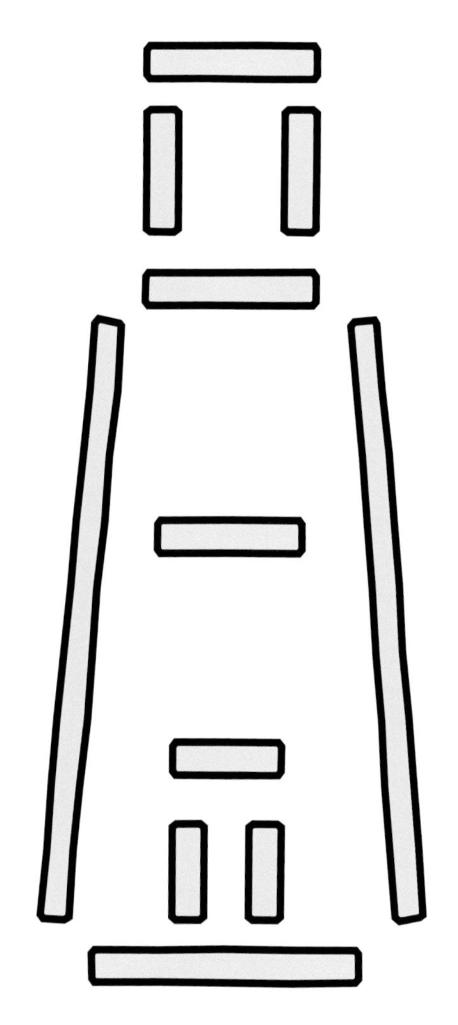

A Lighthouse

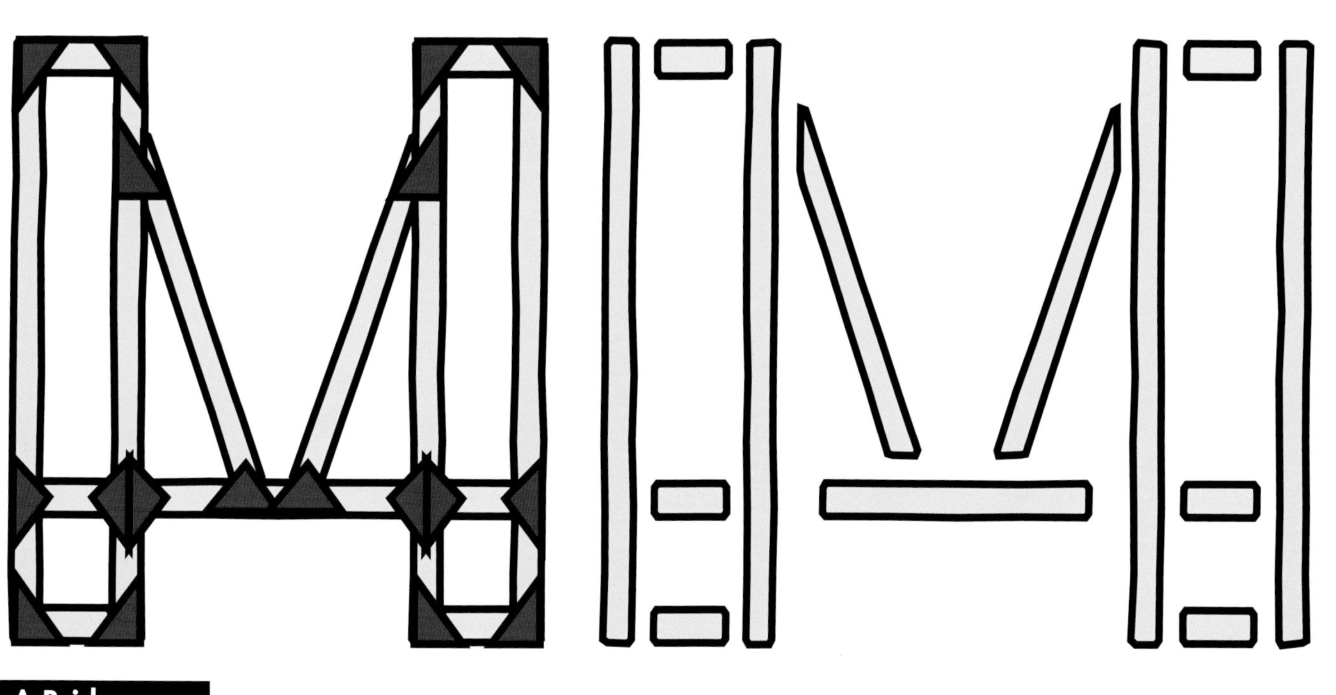

A Bridge

Joining 3D Frames at Corners

'Jinks Technology' affords children the opportunity to build three-dimensional structures that are stable and strong. To change a two-dimensional structure to three dimensions just requires the use of double constructional triangles.

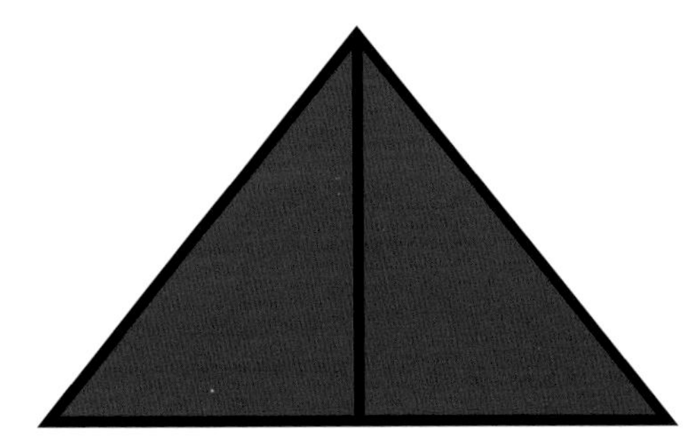

Double constructional triangle

Joining frames at the corners

Most square or oblong 3D structures are produced by joining 2D frames at their corners by using four or more upright supports as shown above.

The constructional triangle used for this type of work resembles the double sized triangle used for axle holders but it is made out of the **thin** card used for ordinary constructional work.

You will need eight of these triangles to join two flat frames.

First, fold the triangle exactly in half along the pencil line to make a right angle. Make a very sharp crease down the fold line. This folded triangle is glued into place at one corner of the base or bottom frame.

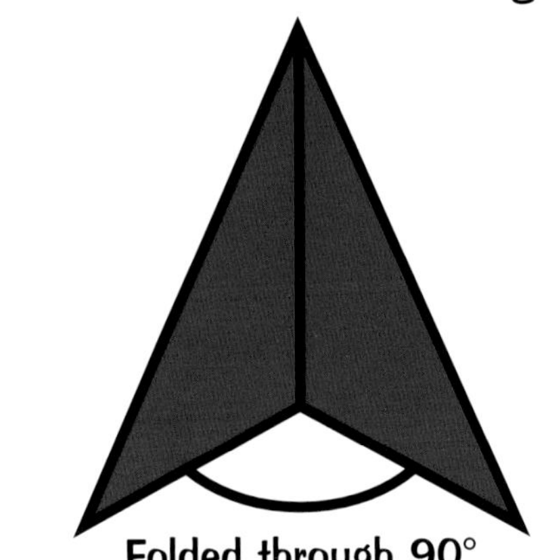

Folded through 90°

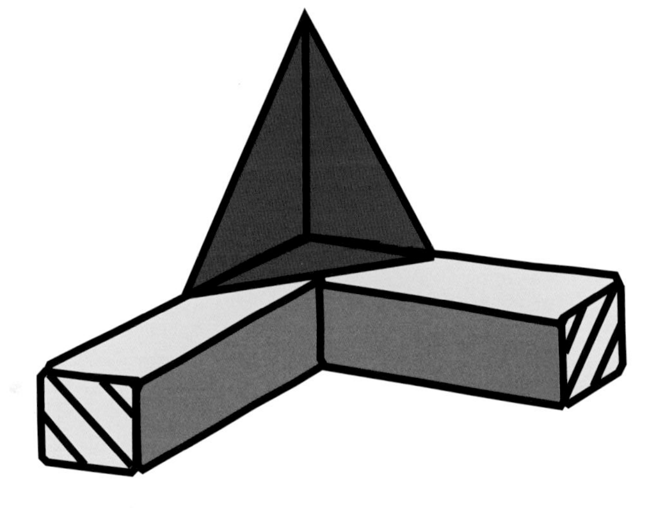

Glued into position to allow 3D building.

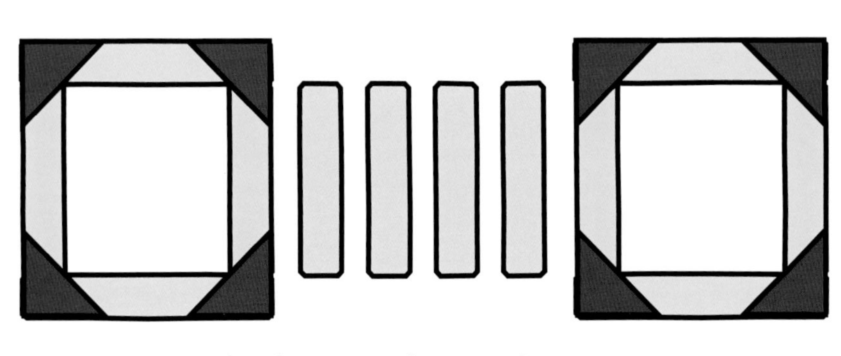

2 Flat frames and 4 upright supports.

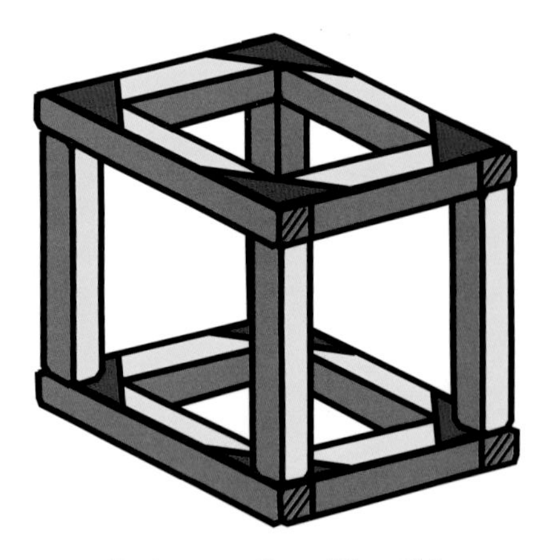

Put together like this.

Three other triangles are similarly glued into position on the other corners of the bottom frame.

Great care must be taken when marking, cutting and folding these corner triangles as the 'squareness' and strength of the finished structure depends on how well these triangles fit into place on the frames and upright supports. Repeat this for the top frame.

PVA glue is then applied to the bottom of one of the upright supports and to the inside of one of the corner triangles on the bottom frame. The support is then pressed into position. Repeat this for the other three upright supports.

Take great care to ensure that the triangle is pressed onto the upright supports properly and that the supports are as vertical and parallel as possible.

Finally, PVA glue is applied to the inside of all of the corner triangles on the top frame and the whole frame is carefully lowered onto the four upright supports. This can be a difficult operation and it often helps to work with a friend. Take care to press all the triangles properly onto the upright supports and then view the whole structure from different angles adjusting it until it is 'square' when viewed from all sides. Remember that this must be done while the PVA glue is still wet. Once dry, it will be difficult to repair a structure that leans to one side.

Joining Different Shaped Frames

Sometimes, particularly when joining shaped frames, it is necessary to join from a middle section of a frame. This is simply done by constructing an upright 'T' joint.

The diagram opposite shows a typical arrangement. Here, a short and long frame need to be joined together.

Where the corners align, corner joints have been used. Where the longer frame overlaps, 'T' joints have been used.

Depending upon the design problem you are trying to solve, one or more of the joints already outlined may usually be used to join the frames together.

Try to select the joint which will give your design the greatest strength.

Remember that the strength of the final structure totally depends upon the strength of the joints that have been used to assemble it.

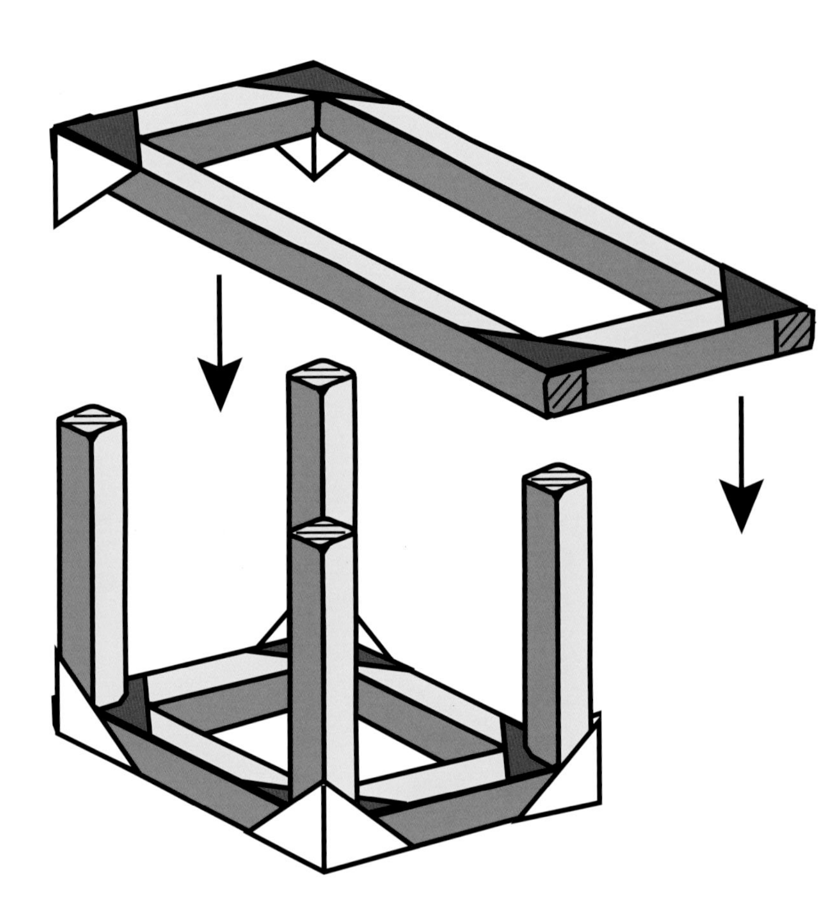

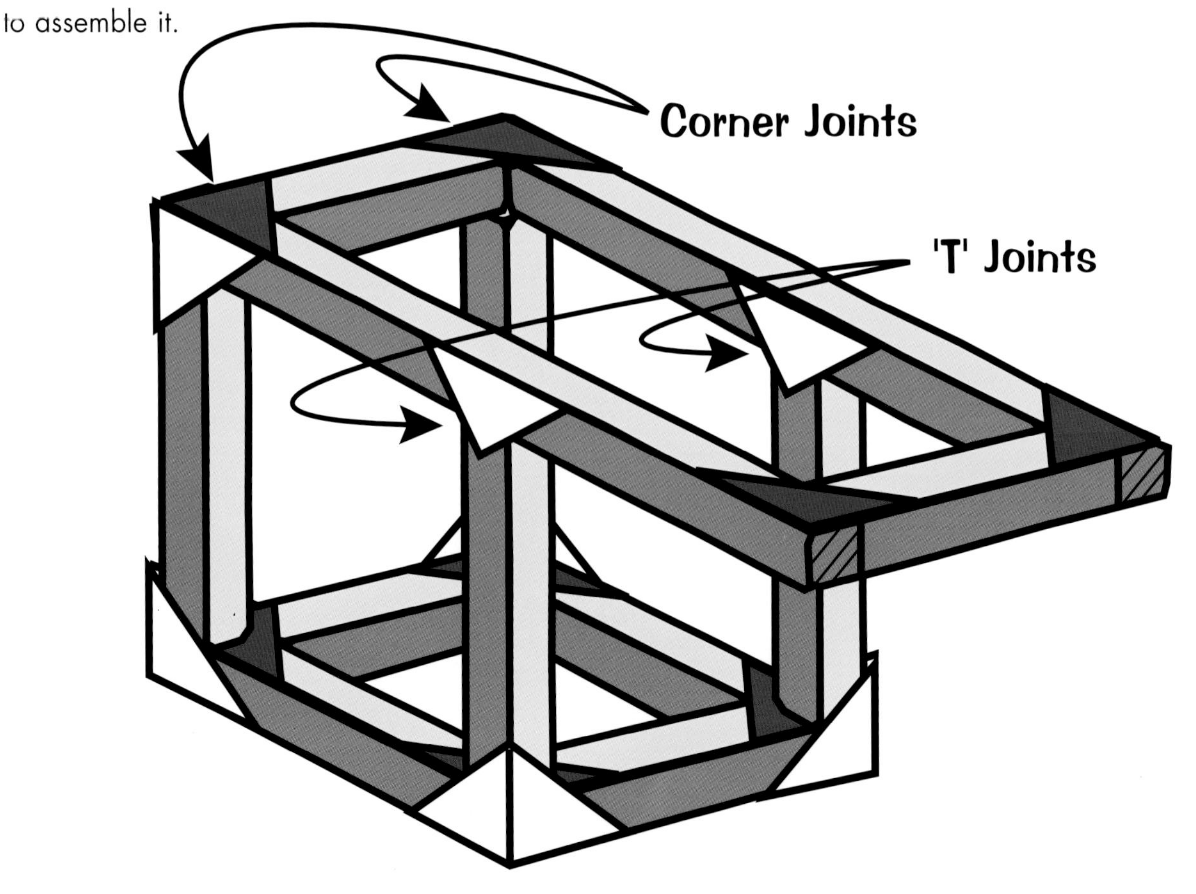

It is a small step from using lollysticks and thin card to create wooden artifacts, to using 8mm x 8mm and /or 21mm x 8mm strips of softwood and thin card to broaden the range of three dimensional outcomes young children can make. The technique is identical but, for added strength, PVA adhesive is applied to the edges of the strips to ensure a solid 'board' is produced. You can make many things using strip softwood.

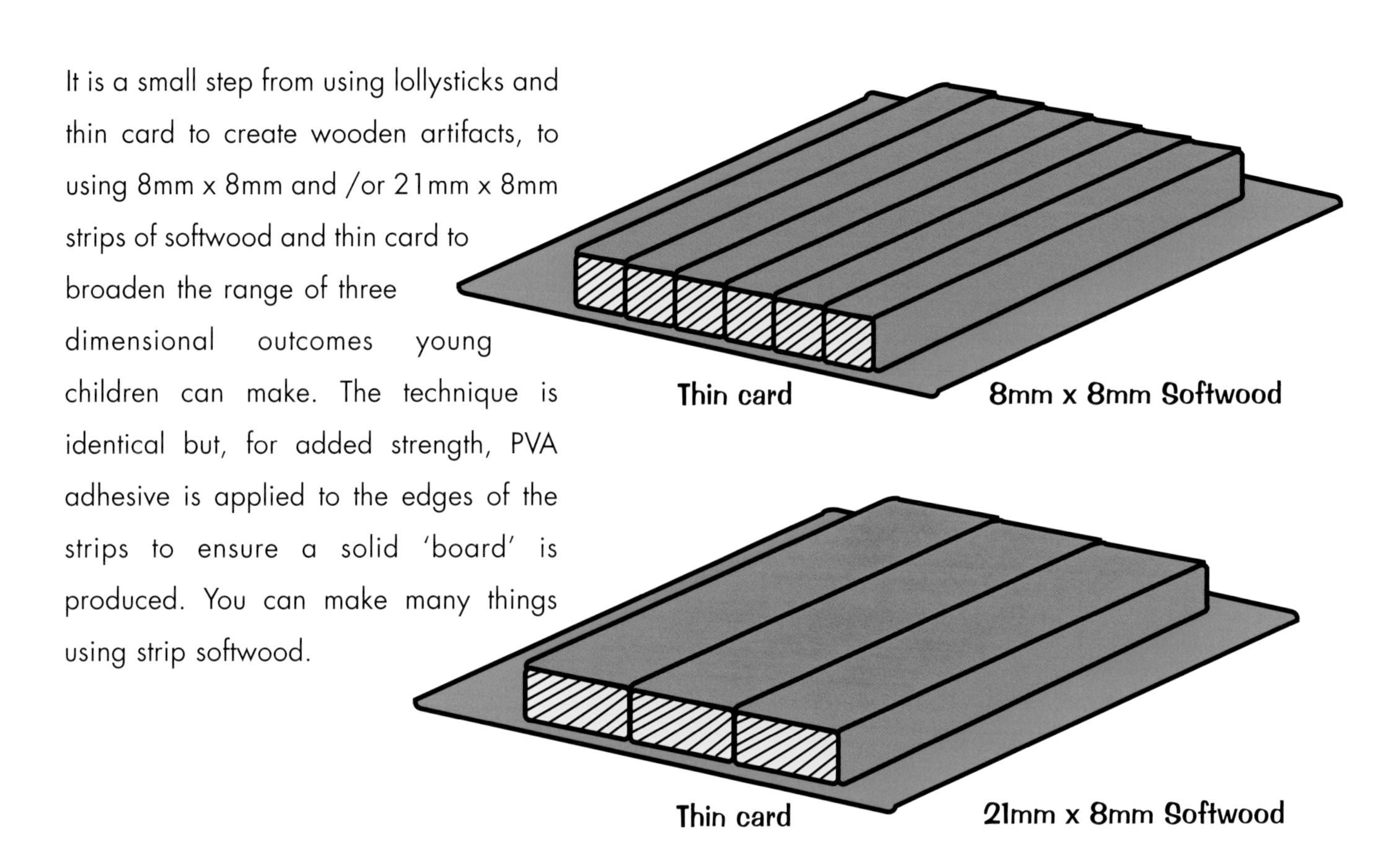

Axle Holder Hinges

Hinges allow a piece of wood or a complete frame to pivot. Hinges can be made in many ways. There are some ideas for hinges shown on this page.

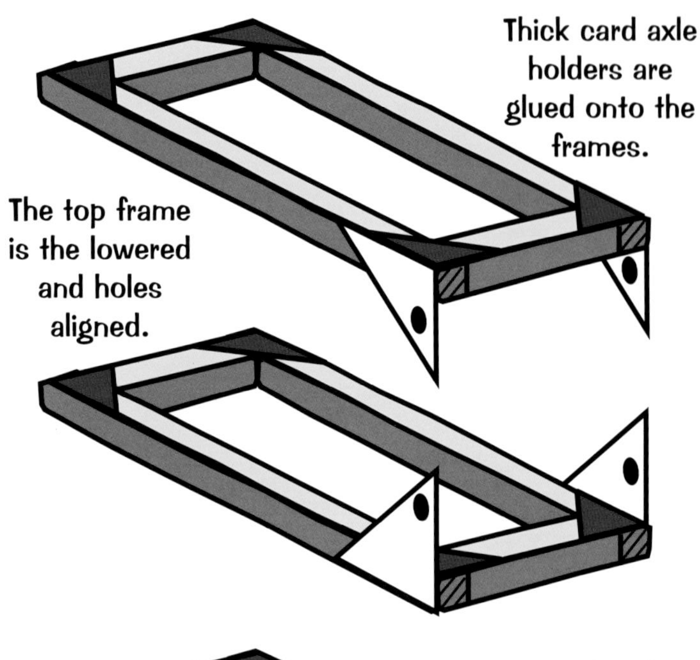

Axle holder hinges

Excellent, free moving hinges may be made by using half or full size axle holders and a 5mm wooden axle.

The holders are glued into position on both frames so that the holes will eventually line up.

Some 5mm plastic tubing should be cut to size to act as a central spacer and a locking mechanism to keep the wooden axle in place.

The central spacer stops the card triangles from being damaged by sideways movement.

There are many variations on this type of hinge depending upon the problem you are trying to solve. Single pieces of wood may be similarly hinged.

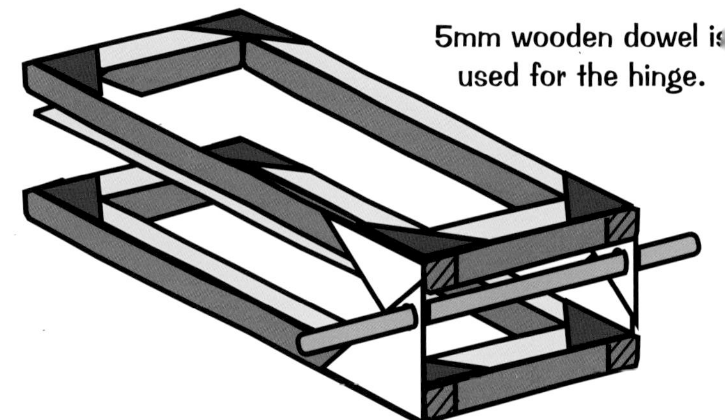

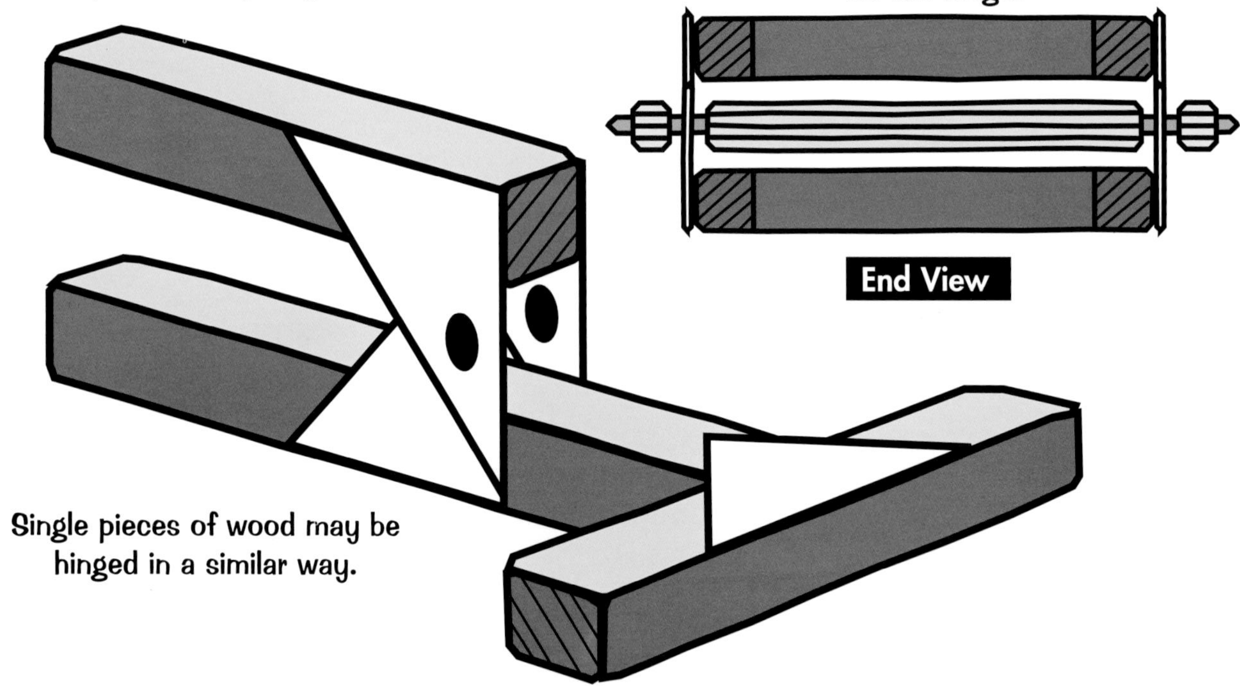

There are times when it is useful to be able to articulate (join but allow them to move) frames and structures. There are many ways to do this depending upon the final strength required. One way is shown below.

Take the two frames or structures to be articulated and glue double size, thick card axle holders into place as shown.

For very light frames, use a single axle holder. For heavier frames use two axle holders, one above and one below the frame.

Bring the two frames together so that the holes in the axle holders align with each other and lock them in place with a piece of 5mm wooden dowel. The dowel can either be held in place with two small pieces of 5mm PVC tubing or for greater strength, two small pieces of square section wood can be drilled with a 5 mm hole and pressed onto the dowel above and below the four axle holders.

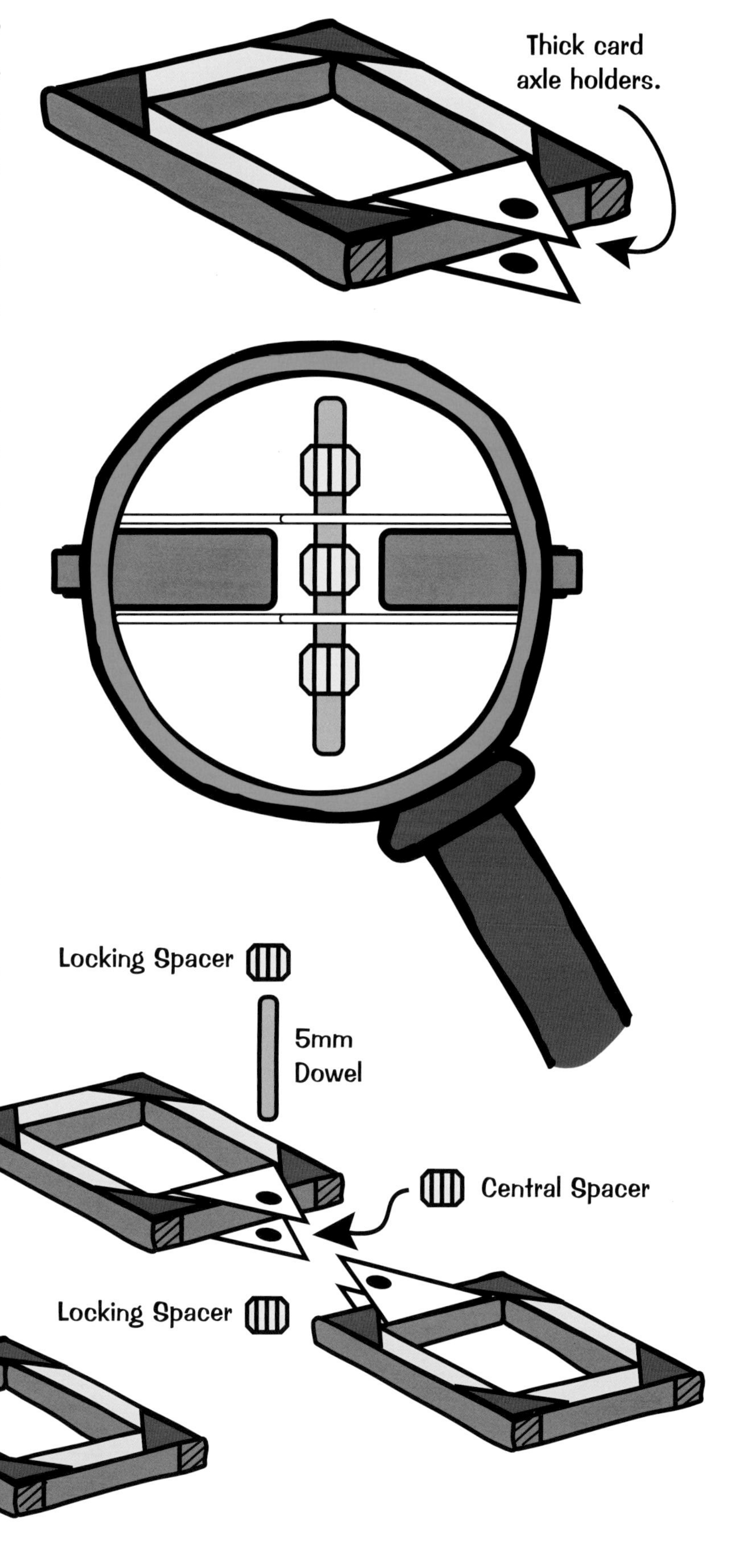

The finished Articulated Joint. This can be used for lorries and trains.

Rotating Frames

There are times when it is useful for one frame or structure to rotate on top of another frame or structure. This is another type of articulation. One way to do this is shown opposite and below.

The method used will vary depending upon the weight of the frames and structures used. For lighter structures, card discs can be used. For greater stability, larger diameter discs can be cut from thick card using a compass cutter. The frames or structures have extra cross supports inserted using 'T' joints and a card or wooden disc (with a 5mm hole to give a loose fit) is glued into place with PVA adhesive above and below the supports. Make sure that the two holes are aligned.

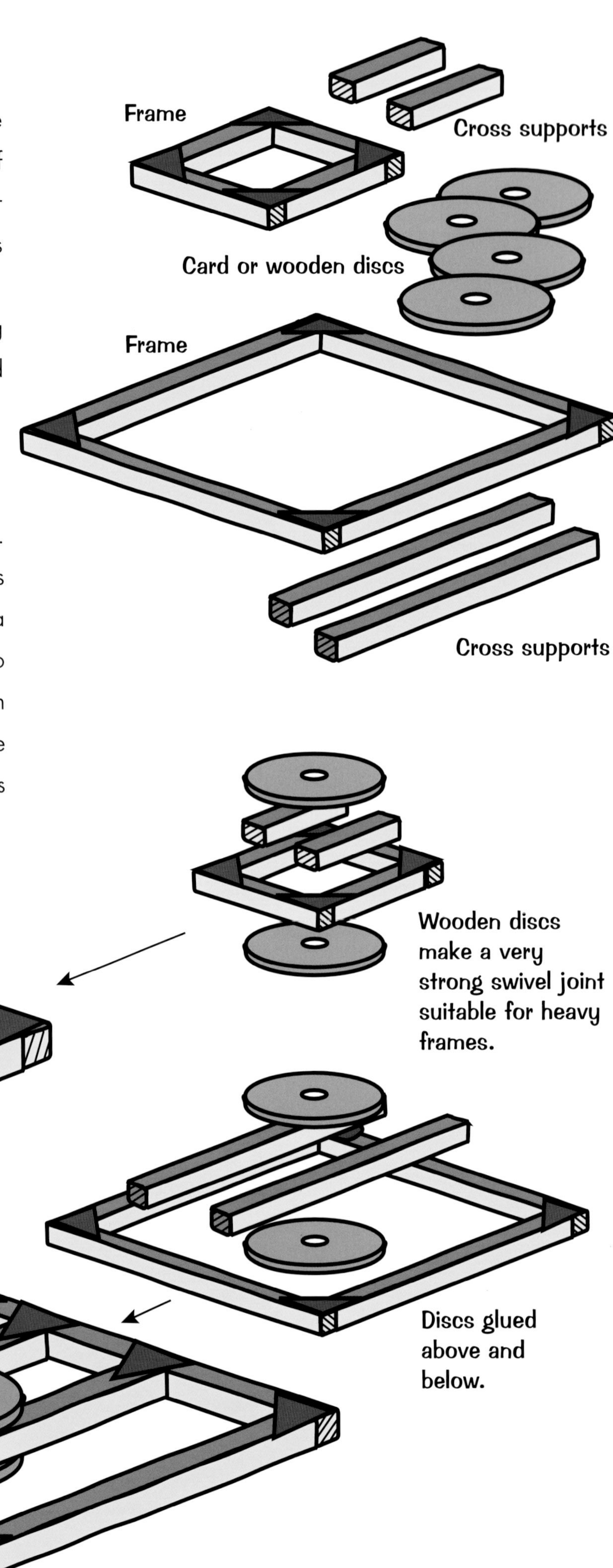

The smoother the disc the better for this type of swivel joint. The top frame sits on the bottom frame and the disc bearings allow the two frames to swivel.

Again, the locking mechanism will vary depending upon the overall strength required. PVC spacers are shown here and they are certainly adequate for light structures.

When building a heavier structure onto the top frame, however (e.g. a crane), a more suitable locking method will have to be used.

Square section wood or a small wooden wheel, drilled with a 5mm hole will lock the 5mm dowel very firmly into place. In these diagrams, only frames are shown, but it would be more usual to swivel complete structures. The frames shown represent the top and bottom frames of the two complete structures

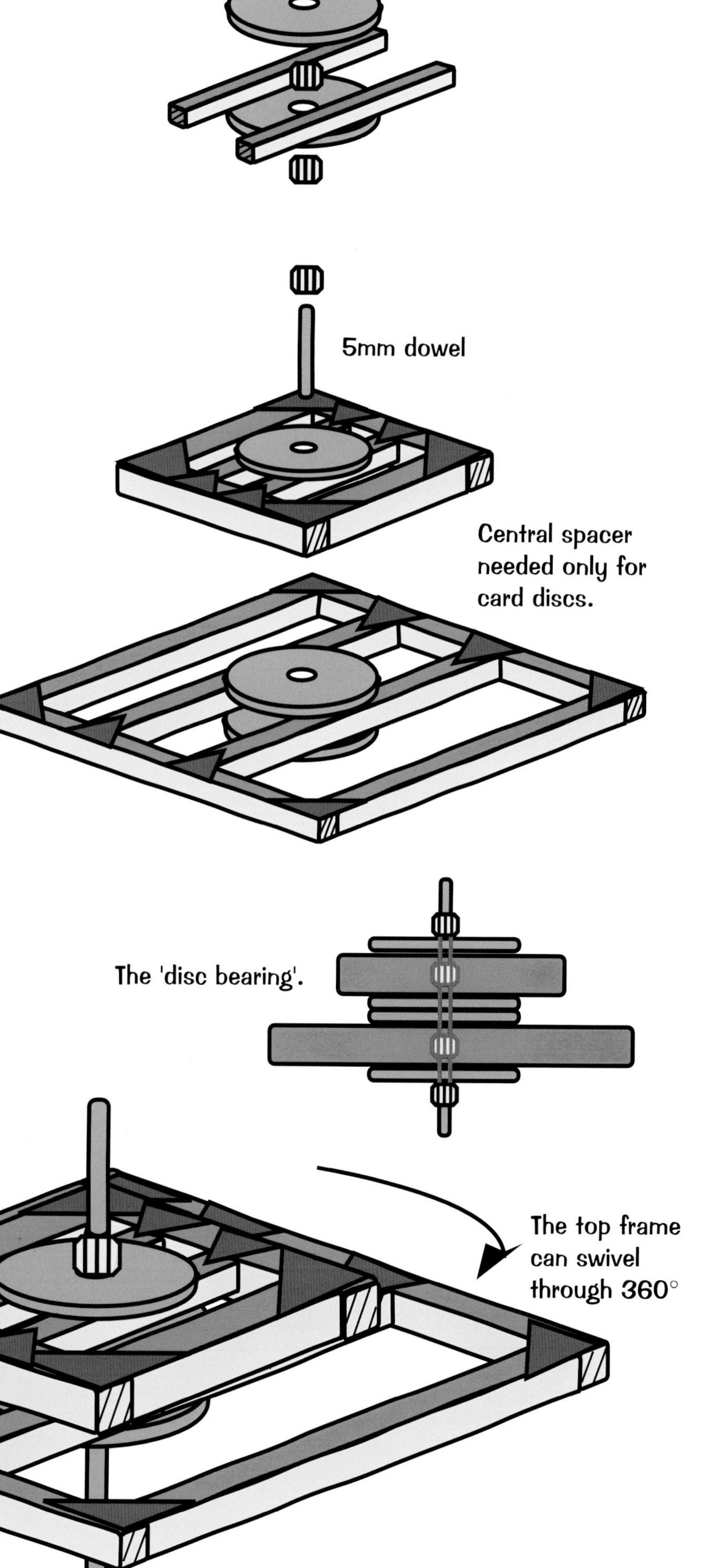

The finished 'Swivel Joint'

Junior Hacksaws

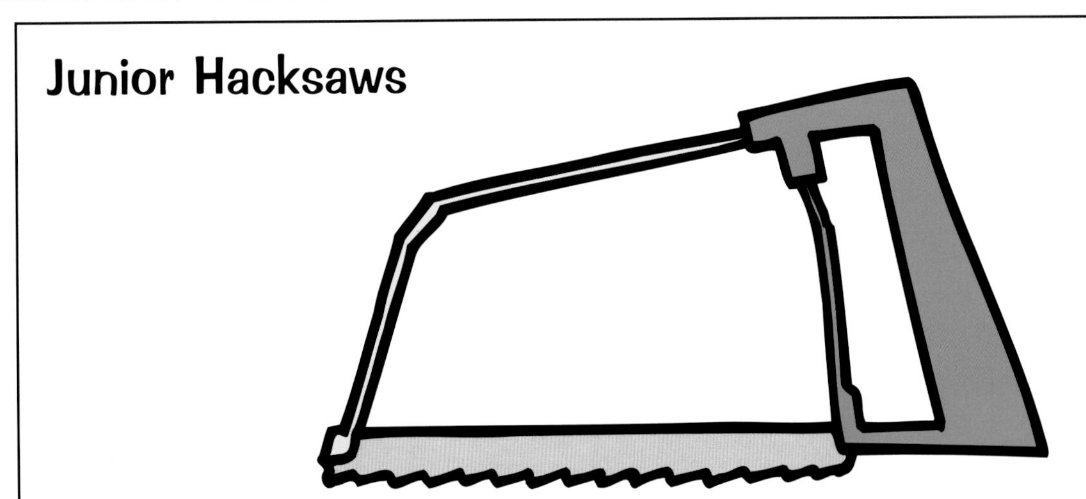

Description

A saw is used to cut small-section wood to a marked size. Junior hacksaws are ideal for this. The saw should always be used with a bench hook. Hold the saw lightly but firmly and keep your eye over the back of the saw. Start your cut by pulling the saw backwards several times to make a small groove in the wood. Don't rush and use the whole length of the blade. Do not press too hard. Let the blade do the work.

Safety

Do not press down too hard. Generally, a sharp saw is safer than a blunt one since you do not have to press as hard when sawing. If the blade of your saw is blunt, tell your teacher.

Replacement Blades

Description

One of the advantages of using junior hacksaws is that you can replace the blade when it gets blunt. The new blade, however, must be inserted correctly. This diagram shows the angle of the teeth. For the saw to work properly the blade must be inserted so that the arrow is pointing **away** from the saw handle.

Safety

Your saw is more likely to slip if the blade is facing the wrong way round so always check it before use. Ask your teacher to help in changing the saw blade.

Single Hole Punch

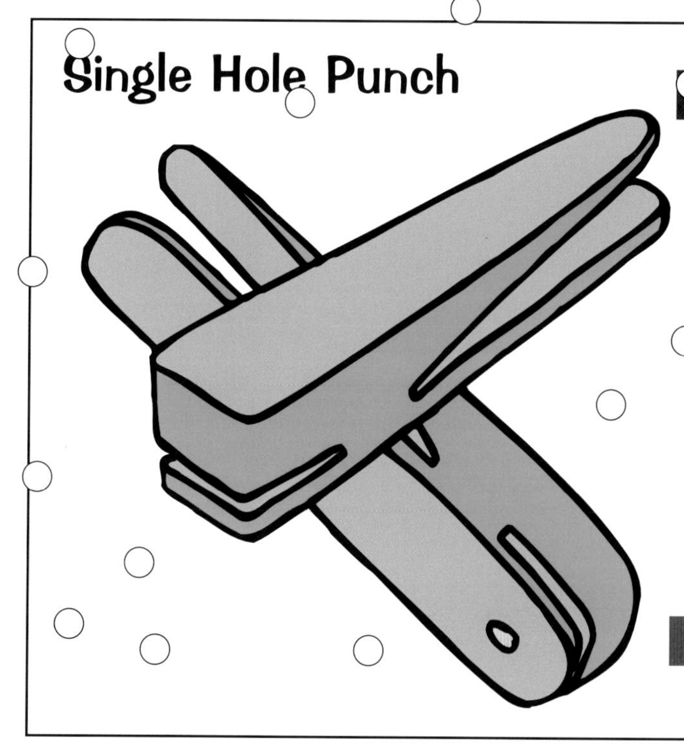

Description

The single hole punch is used to punch neat holes into thick card axle holders, adjustable gear mounts etc. The holes allow 5mm wooden dowel axles to spin freely within the axle holder. Other mechanisms such as wheels, gears, etc. can then be attached to the axle. It is easier to use the punch upside down with the collection tray removed. You can then look through the bottom hole of the punch and accurately place the hole in a particular pre-marked position.

Safety

Like any other tool use it properly.

Bench Hook

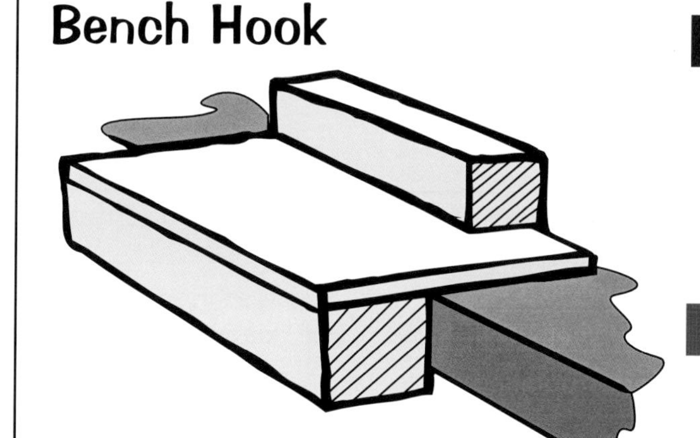

Description

A bench hook helps you to hold a piece of wood firmly while you are sawing it. Hold the wood against the top lip and press the bottom lip against your desk. Saw near to the striped wood.

Safety

When cutting a piece of wood in two, hold the longer piece against the bench hook and saw off the shorter piece.

Centimetre Rule

Description

Centimetre rules have many uses. As well as being used to accurately mark wood to size before sawing, they are also used as a guide for quickly making a grid of squares to produce constructional triangles and axle holders.

Safety

Like any other tool - use it properly.

Pencil Sharpener

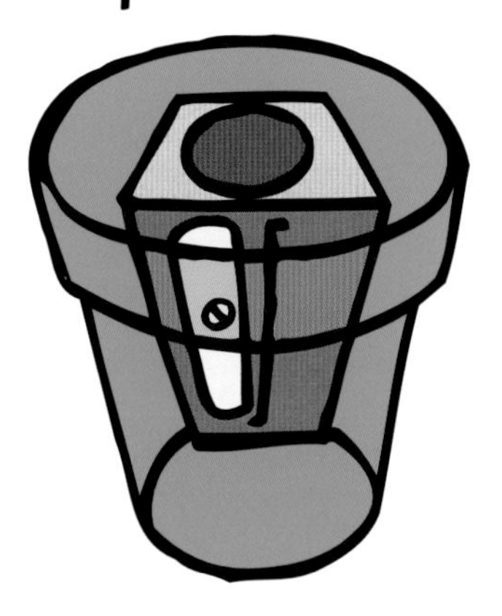

Description

As well as keeping your pencils sharp, a pencil sharpener can be used to slightly taper the ends of 5mm wooden dowel axles. Putting a slight taper on axles helps you to push them into 5mm holes in wheels etc. for a tight "force fit".

Safety

You should try to keep your work area as tidy as possible, so a sharpener with a built in collection box for the shavings is best.

Scissors

Description

Scissors are used for cutting card and paper. Push the card right up to the centre pivot. Don't just use the end of the blades. One long cut will be smoother than several short ones.

Safety

Try not to carry scissors about but if you must, carry them with the handles pointing up and the blades pointing down.

Combination Square

Description

A combination square, as the name implies, is a versatile tool which can do many jobs. One of its main uses is to mark wood prior to sawing at precise angles of 90 and 45 degrees.

Safety

Like any other tool ... use it properly.

Table Vice

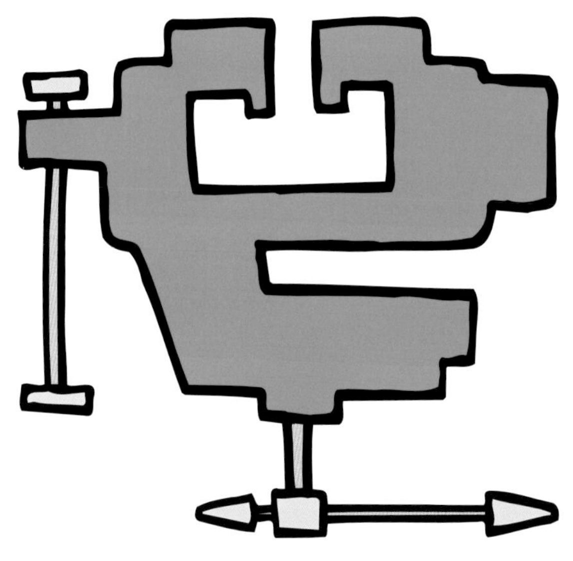

Description

This is a heavy duty holding device which can be attached to a table and used to hold materials steady while you work on them. The jaws adjust to hold materials of different thicknesses and an adjustable mounting device allows the vice to be clamped to tables and desks of various thicknesses. Leave the jaws of the vice very slightly open when not in use.

Safety

A vice can exert enough pressure to squash fingers and break bones so never play with the jaws of a vice and be careful as you lock the material into place. Never turn run near a vice. It is very hard and can hurt you if you fall against it.

G-Clamp

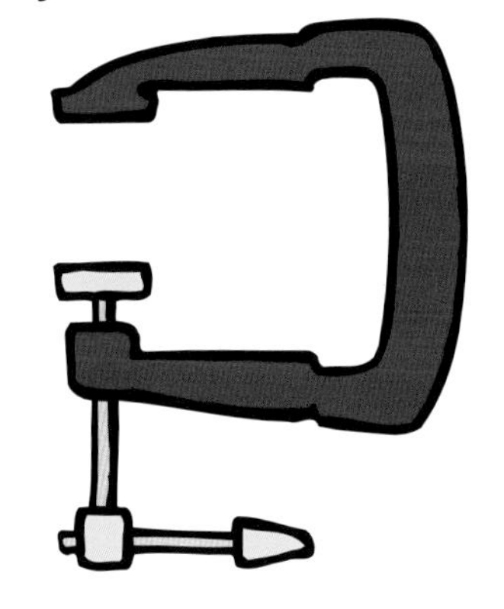

Description

A G - clamp is a medium duty holding device which allows you to hold materials very firmly together e.g. after gluing, or to hold materials temporarily to a table or desk.

Safety

Like the vice, G - clamps can exert extremely high pressure when fully tightened so never play about with them. Store them away safely when not in use.

Hammer

Description

A hammer has many uses e.g. driving a nail punch into a syrup tin lid, gently tapping tight dowel into a 5mm hole etc. Balance the hammer lightly but firmly in your hand and hammer gently. Always hold the tool or material that you are striking very firmly.

Safety

The greatest danger with a hammer is that the hammer itself or the tool/material that you are striking will slip. Hold the material firmly and be careful not to hit your own fingers.

Craft Knife*

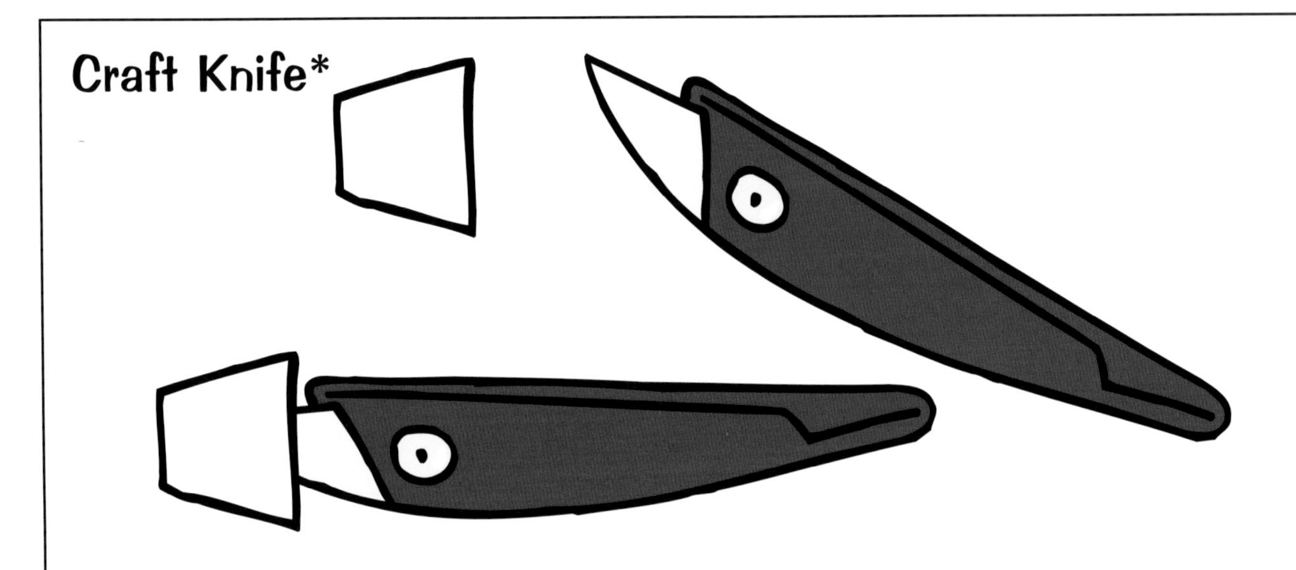

Description

Craft Knife and Safety Rule. A craft knife must: **ALWAYS BE USED WITH A METAL SAFETY RULE.** It is used to make long straight cuts in card or paper or to cut shapes from inverted boxes before gluing them back together.

Right handers, hold the knife in your right hand and use the safety rule with your left. If you are left handed, use your left hand for the knife.

Hold the rule very firmly, keeping your fingers inside the indent. Hold the knife lightly but firmly against the metal edge of the safety rule and draw the knife towards you.

Don't press too hard. If the knife does not cut through the material in one go repeat the cutting action until it does. Thick card may well take several strokes of the knife before it is cut all the way through.

Safety

Craft knives can be **VERY DANGEROUS** if you misuse them. Concentrate hard when you are using the knife and **NEVER** use a craft knife without a metal safety rule. Like the saw, a sharp knife is safer than a blunt one since you do not have to press down as hard. If you think the blade of your craft knife is blunt, tell your teacher straight away. If your knife has a safety cover, replace it as soon as you have finished using the knife. A cork can be used as a safety cover as shown in the drawing. Carefully push the blade of the knife into the cork when you have finished using it.

NEVER walk around with a craft knife unless the blade is covered.

Safety Goggles

Description

Safety goggles should always be worn whenever there is a danger of a tool or material coming into contact with your eyes.

Safety

Don't pull the strap too tight!

Nail Punch

Description

A nail punch is used with a hammer to make holes in metal, syrup tin rollers so that 5mm axles can pass through the holes. Start the hole with a solid knock from the hammer then continue to hammer lightly, turning the punch between blows. Keep trying the axle in the hole. Remember, you can always widen the hole but you can't make it smaller.

Safety

The greatest danger here, is that the punch might slip when you first start off the hole. Place some masking tape over the point where you wish to make your hole. The punch is much less likely to slip on the tape.

Round File

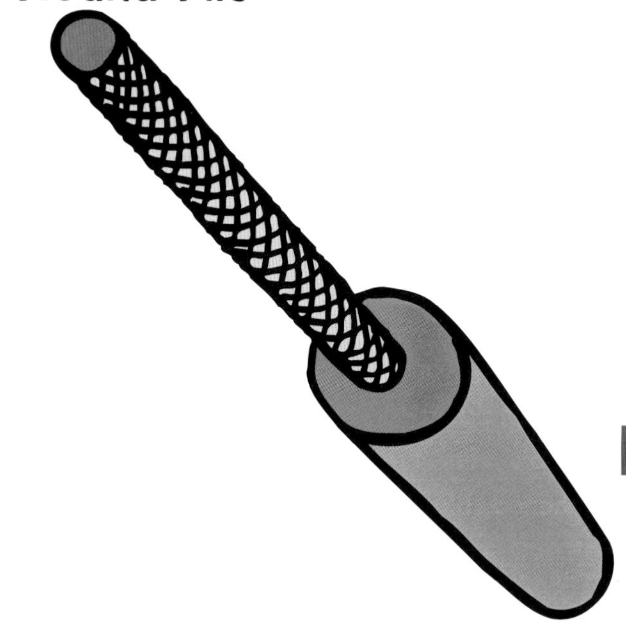

Description

Round files can be used to remove wood and other material. You would normally drill a 5mm hole in a wooden wheel to make it a tight "force fit" on a 5mm wooden axle. Sometimes, however, particularly on ready-made wheels, the hole might be a little too tight. You can enlarge the diameter of the hole slightly with a round file but take care not to remove too much material.

Safety

The material you are working with must be held firmly when using a file. It is often useful to hold the material in a vice, then you should hold the file with both hands and don't press too hard.

Cutting Mat

Description

A cutting mat should always be used underneath your work when you use sharp tools such as craft knives and compass cutters. This protects the surface of the table or desk that you are working on and also the sharpness of the cutting blade.

PVA Glue pot & Spreader

Description

For most of your work you will use PVA adhesive which should be kept in an airtight pot. Old 35mm film containers are ideal for this.

A spreader allows you to spread the glue evenly onto wood and card and stops the PVA from coming into contact with your skin. Remember to remove all the PVA from the spreader at the end of the lesson. This should wash off in water.

If you get PVA on your clothes, sponge it off immediately with a damp cloth. Use lots of fresh cold water. Once PVA has dried, it is very hard to remove from clothing.

Safety

PVA glue is quite safe if used properly but you must avoid contact with your eyes so wash your hands frequently and try not to rub your eyes. If you do get any glue in your eye, tell your teacher immediately and wash your eye with lots of cold water. PVA can cause slight irritation to sensitive skin so always use a spreader.

Pliers

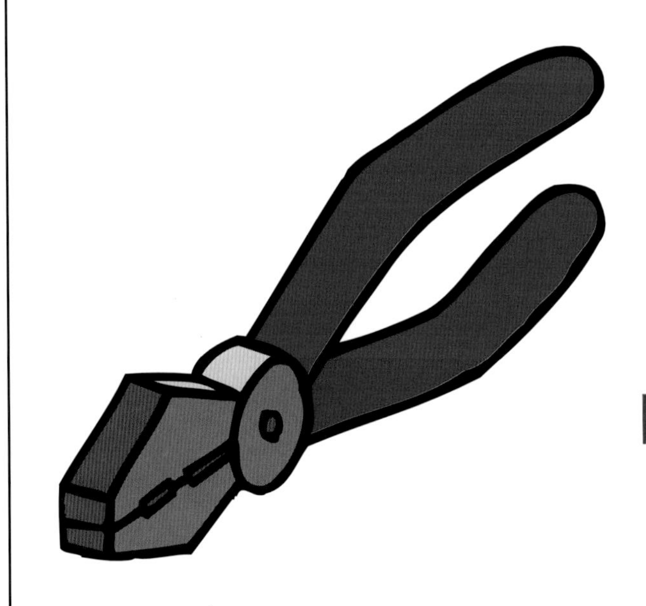

Description

Pliers are another general purpose tool used for both cutting and holding. The jaws have a serrated holding edge which can grip a variety of materials very firmly. They also have a short, toughened blade which can cut through hard materials such as coat-hanger wire etc.

The nearer you place the material to the pivot or the further from the pivot you grip the handles, the stronger the holding or cutting force.

Safety

Because of their design, pliers exert tremendous holding and cutting forces so keep your fingers well away from their jaws. If you are using them to cut through a tough material, wear safety glasses as the material can be thrown out with great force when cut.

Hand Drill and Stand

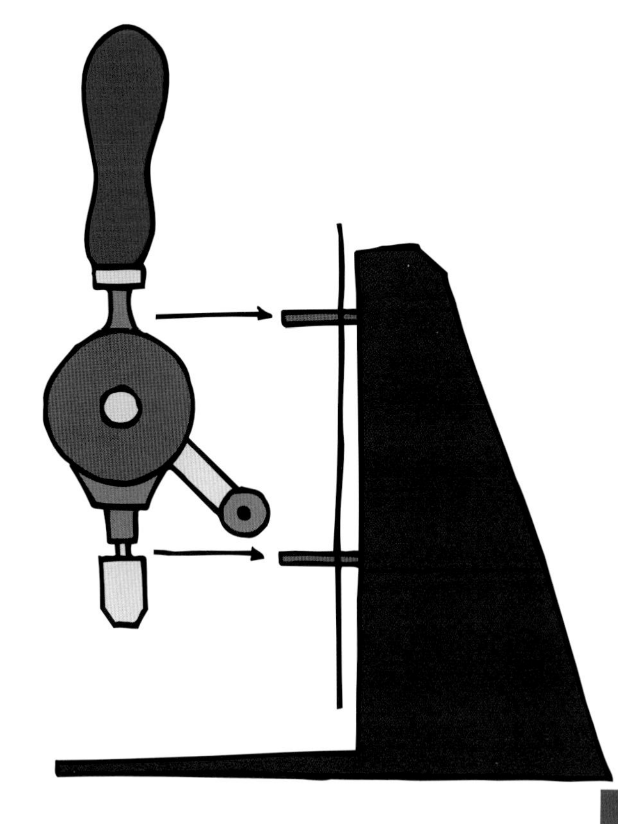

Description

A hand drill will allow you to drill holes in wood. Wooden wheels, pulleys etc. have to be drilled so that they can be fitted onto 5mm wooden dowel. The adjustable chuck at the end of the drill allows different sized bits to be fitted. The spiral pattern on the bit removes the sawdust from the hole. A 5mm hole will give a tight fit on a 5mm axle. Smaller sizes can be used for pilot holes. A drill stand will keep the drill straight and is adjustable for right and left handed use. Use one hand on the upright handle to press lightly down onto the material to be drilled. Use the other hand to turn the drill mechanism clockwise. Place the material to be drilled onto a piece of scrap wood and get a friend to hold it very firmly while you control the drill. Don't press too hard and like the saw, let the twist drill do the work. Drill right into the scrap wood. To remove the drill, keep turning the handle clockwise, lifting the drill at the same time.

Safety

Drills, especially in a drill stand are quite safe if used properly.

Like the saw and the craft knife, however, you must not press too hard. If the twist drill bit is kept sharp you will not have to press hard and most materials can be drilled very quickly. If you press too hard, thin materials such as lolly sticks will split and there is the danger that the drill might slip out of the clips in the drill stand and injure the person holding the material for you.

Take your time and let the tool do the work.

Pencil

Description

A pencil will mainly be used with a centimetre rule to mark a length of square section wood or 5mm dowel before sawing to size. If you are to cut your wood accurately, you must keep your pencil sharp.

Safety

Sharp pencils can be very dangerous, always put your pencil down when you are not using it.

Compass Cutter*

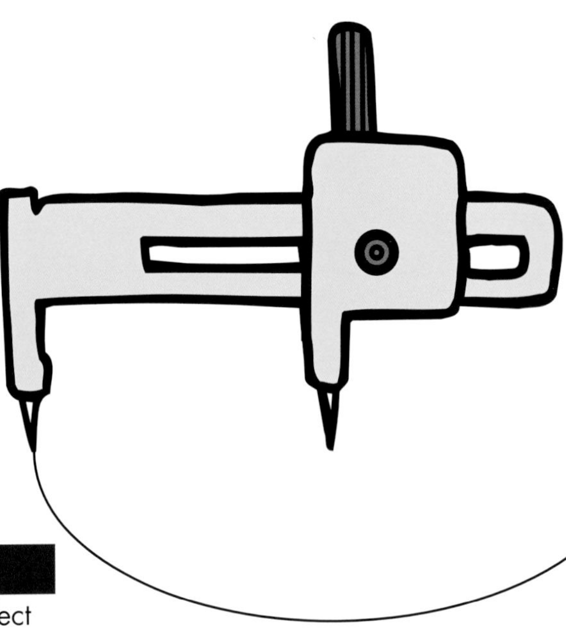

Description

A compass cutter allows you to cut perfect circles from a variety of materials e.g. thin and thick card, paper, cloth etc. It is a type of adjustable compass but with a craft knife blade replacing the usual pencil. It can, therefore, cut circles of different diameters.

A compass cutter must always be used on some kind of cutting mat. This protects the table or desk from the sharp craft knife blade as well as giving the point of the compass cutter something to bite into so that it doesn't slip. Like the craft knife, never press too hard with the compass cutter. If you don't cut through the material in one go, continue turning the cutter around in a circle going over the cut line.

Safety

The compass cutter carries many of the same dangers as the craft knife since it has a similar blade. Because of the way it is used, however, it is usually less accident-prone.

Avoid carrying it around unnecessarily and protect the sharp blade when not in use.

Never try to force the blade through a material or the point of the compass cutter could jump out of its hole and the blade could then cut your hand.

Glass Paper

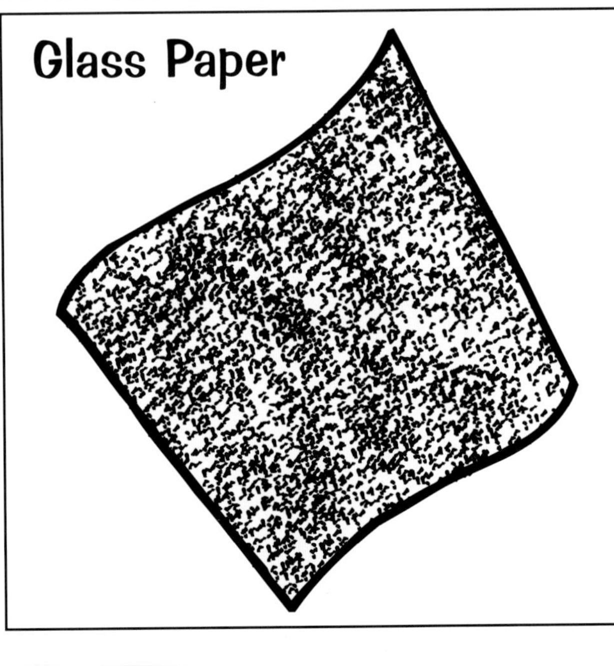

Description

Glass-paper is used to smooth off wood and other materials. Whenever wood is sawn, there is always a rough edge at the bottom of the cut. This can be smoothed down with glass-paper before joining. If 5mm wooden dowel is too tight a fit into PVC tubing, wheel etc., its diameter can be reduced slightly by rubbing with glass-paper. To keep its roundness, wrap the glass-paper right around the dowel and use one hand to move the glass-paper quickly up and down. While doing this, keep turning the dowel with your other hand.

Glossary of Terms & Bibliography

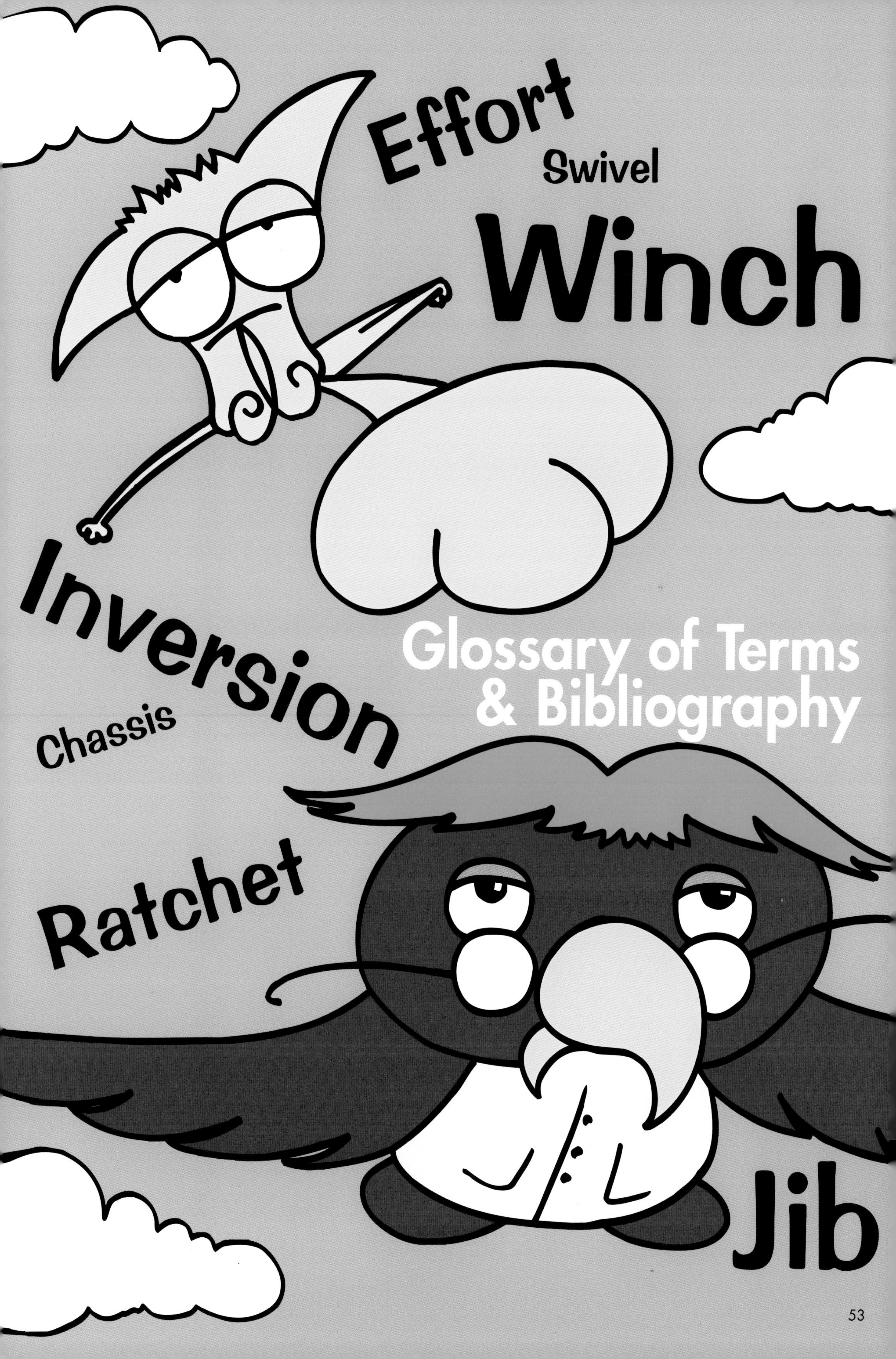

Glossary Of Terms for Structures and Mechanisms

Align	To line two or more things up with each other in a set pattern. To align axles on a vehicle is to make sure that the front and back axles are parallel to each other and to the vehicle itself.
Axle	A rod/shaft on which a wheel rotates. The wheel/s can either run freely on the axle or be fastened to, and turn with, the axle.
Axle holders	Triangular pieces of thick card, punched with a 5mm hole, through which an axle fits and rotates.
Bearing	The part of a machine that bears friction.
Belt drive	Pulley systems using a belt to transmit motion and force from the driver pulley to the driven pulley.
Block and tackle	An arrangement of pulleys and cables to make the lifting and lowering of very heavy loads easier.
Cam	A device used to change rotary motion into either reciprocating or oscillating motion.
Chassis	The frame upon which a vehicle is built.
Circumference	The distance around a circle.
Compression	A force that tries to squash a structure.
Correx/Corriflute	A plastic material very similar in appearance to corrugated paper or card.
Compass Cutter	A compass to cut circles, or parts of circles, in paper or card.
Crane	A machine for lifting and lowering loads. These machines are often used at building sites, docks etc.
Design	'To design is always to provide some form, structure, pattern or arrangement for a proposed thing, system or event'. Keith-Lucas Rep. 1980.
Diameter	The distance across a circle passing through its centre.
Dowel	Lengths of hardwood, circular in cross section, used for axles or as wooden pins.
Driver gear	A gear which drives other gears. This gear is, itself, usually driven by some other mechanism, winder, electric motor etc.
Driven gear	A gear which is driven or turned by a driver gear.

Drive ratio — Drive ratio applies to both gears and belt drive mechanisms and is the difference between the number of teeth (gear) or the circumferences/ diameters (belts) of the driver and driven gear/pulley; the drive ratio determines how rotary motion will be speeded up or slowed down by a system of gears or belt driven pulleys.

Effort — The energy we put into machines to produce work.

Follower — A device which follows the movement of a cam. The way it moves and the distance it moves depends on the shape of the cam.

Force fit — A tight fit between a wheel and axle, usually requiring the wheel to be twisted or tapped onto the axle using a pin hammer.

Frame structure — Frame structures are made from strips of material joined together to form a framework e.g. a crane, a pylon etc.

Friction — Force of resistance which is encountered when two things rub together or move over one another.

Fulcrum/Pivot — The fixed point upon which a lever moves.

Gear wheel — A wheel with teeth around its circumference.

Gear train — The name given to a number of gears, in a line, connected together

Guide — A device that determines the direction of movement of string/rope/cable etc.

Hardboard — Compressed wood fibres and glue in thin sheet form (usually 3mm).

Impact Adhesive — A glue used to stick non-porous materials such as plastic. It is applied to both surfaces and allowed to dry before the surfaces are brought together - the surfaces stick on impact. It is now available in a water-based form but either type must always be used in a well ventilated area because of the danger of fumes. Only to be used under direct teacher supervision.

Inversion — To turn in or about; to reverse.

Jib — The inclined arm of a crane.

Jig — An appliance for guiding a tool.

Load That which is to be lifted/carried/moved etc. by the application of an effort.

Lever Levers consist of a beam that can rotate about a fixed point called a fulcrum or pivot. In use, levers allow a load to be moved when effort is applied. There are three main types of lever:

Lever 1st Class: here the pivot is in the middle with the effort being applied on one side and the load on the other. Examples of a first-class lever would be a pair of scissors, a pair of pliers, a child's seesaw, oars in a rowing boat.

Lever 2nd Class: here the pivot is at one side with the effort being applied at the other side. The load is in the middle. Examples of a second-class lever would be nutcrackers, a wheelbarrow, a car foot pump.

Lever 3rd Class: here the pivot is at one side and the effort is applied in the middle, the load is at the other side. Examples of a third-class lever would be a pair of tweezers, a fishing rod, a garden spade, a mangonel (siege weapon).

Mechanism A way of turning one kind of force into another kind of force - the parts of a machine that enable it to perform its function.

Mesh The engagement of geared wheels.

Needle file Very small files, in a variety of shapes, for precision work.

Net The shape of a three-dimensional figure when laid out flat.

Pawl A device which locks a ratchet in place.

Pilot hole A small diameter hole usually drilled to accept a screw as a guide for a larger diameter hole.

Pin hammer A 'Warrington' hammer with a very light head (100 grms).

Pulley wheel A wheel with a groove around its circumference.

PVA Glue Polyvinyl Acetate glue is a white liquid which provides strong joints between porous materials e.g. wood, paper, card etc. PVA glue is not usually waterproof though waterproof types are now becoming available.

Ratchet A wheel with inclined teeth with which a Pawl engages.

Roller A very wide wheel which is circular in cross-section, reducing the friction of a vehicle/load etc. being moved.

Shell structure Shell structures are held together by the skin of the material they are made from e.g. a drinks can or carton.

Spacers Small pieces of 5mm internal diameter plastic tubing which act as a type of washer.

Speed How fast a vehicle/person/thing is moving. Average speed is calculated by dividing the distance travelled, by the time taken to travel, it e.g:-
If a vehicle travels 30 centimetres in 10 seconds, we say that its average speed is 3 centimetres per second (cm/sec).
If a vehicle travels 60 metres in 5 seconds, we say that its average speed is 12 metres per second (m/sec) etc.

Structure A structure provides support and must be able to resist forces. A structure must be able to support its own weight and whatever weight or load it is designed to support.

Swivel A link or pivot enabling one part of a structure to revolve without turning another.

Technology 'The art of making things' - Sir Norman Foster.
'The application of curricular, material and human resources to the solution of perceived needs' - S.C.S.S.T/David Jinks

Tension A stretching or pulling force.

Tensioner A device for applying a continuous force of tension to a mechanism.

Vehicle A structure, in or on which persons or goods are transported.

Washer A circular disc of metal/rubber/plastic with a hole in its centre used with nuts and bolts to spread the load or secure the nut.

Wheel Anything which is circular in cross-section which can be attached to an axle to allow the movement of a vehicle - it is a slice of a roller.

Winch A device for winding string/rope/cable in or out.

Winder That part of a winch mechanism which allows you to turn it, usually involving a handle which is offset from the central axle of the winch.

Bibliography

A History of Handicraft Teaching - Blachford, G. - Chatto & Windus - 1961

An Introduction to Craft, Design and Technology - Stewart Dunn - Bell & Hyman - 1986

Be Safe - Association for Science Education - 1990

Craft, Design and Technology from 5 to 16 - D.E.S. - 1987

Craft, Design and Technology - Foundation Course - Finney, M. & Fowler. P. - Collins - 1986

Design Graphics - David Fair and Marilyn Kenny - Hodder and Stoughton - 1987

Design and Primary Education - The Design Council - 1987

Design and Technological Activity - Assessment of Performance Unit - H.M.S.O. - 1987

Design and Technology Through Problem Solving - Robert Johnsey - Simon and Schuster - 1990

Design and Technology 5 to 12 - Pat Williams and David Jinks - Falmer Press - 1985

Design and Technology in the National Curriculum - Draft Proposals - S.C.A.A. - 1994

Design and Technology in the National Curriculum - DfEE - 1995

Design and Technology - The National Curriculum for England - DFEE & Q.C.A. - 1999

Engineering Among the Schools - Page, G. T. - Inst. of Mech. Eng - 1965

Half Our Future - The Newsom Report - H.M.S.O. - 1963

Institute of Handicraft Teachers, Conference Handbook - 1964

Make It Safe - National Association of Advisers & Inspectors In Design and Technology - 1992

Monitoring the School Curriculum: Reporting to Schools - (QCA) - 1997

National Curriculum Technology: the Case for Revising the Order - N.C.C. - 1992

Primary Education In England - D.E.S. - H.M.S.O. - 1978

Problem Solving In Science and Technology - David Rowlands - Hutchinson - 1987

Problem Solving: Science and Technology In Primary Schools - Engineering Council and SCSST -1985

Science for Ages 5 to 16 - Sec. of State for Education Working Party Report - 1988

Technical Subjects In Secondary Schools - Scottish Educ. Dept. - 1950

Technical Teaching and Instruction - McClennon, A. - Oldbourne. - 1963

Technology in the National Curriculum - D.E.S. - 1990

Technology in the National Curriculum - Getting It Right - Alan Smithers & Pamela Robinson - Engineering Council - 1992

Technology programmes of study & attainment targets: recommendations to the Sec. of State" - N.C.C.-'93

The National Curriculum. Handbook for Primary Teachers in England - DfEE & Q.C.A.

The Revised National Curriculum for 2000. What has changed? - Q.C.A. - 1999

The Secondary Technical School - Edwards, R. - Univ. Lon. Press. - 1960

The Way Things Work - Macaulay. D. - Dorlng Kindersley - 1988

Understanding Design and Technology - Assessment of Performance Unit - D.E.S. - 1981

l5 to 18 - The Crowther Report - H.M.S.O. - 1959